Transportation, Land Use and Integration

WIT PRESS

WIT Press publishes leading books in Science and Technology.
Visit our website for the current list of titles.
www.witpress.com

WIT *eLibrary*

Home of the Transactions of the Wessex Institute, the WIT electronic-library provides the international scientific community with immediate and permanent access to individual papers presented at WIT conferences.
Visit the WIT eLibrary at http://library.witpress.com

Transportation, Land Use and Integration

Applications in Developing Countries

Edited by

I.M. Schoeman
North West University, South Africa

WIT*PRESS* **Southampton, Boston**

Editor:

I.M. Schoeman
North West University, South Africa

Published by

WIT Press
Ashurst Lodge, Ashurst, Southampton, SO40 7AA, UK
Tel: 44 (0) 238 029 3223; Fax: 44 (0) 238 029 2853
E-Mail: witpress@witpress.com
http://www.witpress.com

For USA, Canada and Mexico

Computational Mechanics International Inc
25 Bridge Street, Billerica, MA 01821, USA
Tel: 978 667 5841; Fax: 978 667 7582
E-Mail: infousa@witpress.com
http://www.witpress.com

British Library Cataloguing-in-Publication Data

A Catalogue record for this book is available
from the British Library

ISBN: 978-1-78466-233-2
eISBN: 978-1-78466-234-9

Library of Congress Catalog Card Number: 2016958017

Printed by Lightning Source, UK.

About the Authors

Dr Werner Heyns is a technical director in the Pretoria office of Aurecon (Pty) Ltd where he has been specialising in transport planning and traffic engineering. He has been practicing transport planning for nearly 20 years offering a wide skill set from conceptual planning to design to implementation. He has extensive experience in transport development planning helping developers minimise cost associated with transport infrastructure without compromising their sustainability needs. In the last 8 years he has also delivered a variety of national policy plan updates or reviews, notably the Passenger Rail Agency South Africa's National Strategic Plan, the National Department of Transport's Transport White Paper, the National Transport Master Plan 2050 and the National Land Transport Framework. He has also been involved in projects promoting sustainable development, the delivery of transport impact assessments, strategic modelling, feasibility studies, master planning, policy formation studies, green travel plans and parking and traffic studies. Werner has also provided traffic and transport input into Strategic Environmental Assessments and Environmental Impact Assessments for a variety of mining house and environmental clients. Werner is a professionally registered Planner with the South African Council for Planners, a Member of the Institute of Highways and Transportation and a Chartered member of the Institute of Logistics and Transport. In the past decade Werner has published a number of papers on traffic congestion management, demand forecasting and technology choice in public transport and land-use management and transport planning integration. You can contact Werner at werner.heyns@aurecongroup.com.

Mr Simon van Jaarsveld is an associate in the Johannesburg office of Arup (Pty) Ltd. Simon specializes in transport planning and engineering on both local and strategic level. Simon has almost 20 year's experience. After graduating his BSc Hons Town and

Regional Planning carrier at the University of Pretoria in 1996, he started off his career as a town planner. For five years Simon developed strategic planning skills by assisting in the development of integrated development frameworks, environmental studies and provincial road impact studies. During this period, Simon completed his BSc Hons Transportation and Traffic Engineering at the University of Pretoria. Simon took an opportunity to move to Dublin, Ireland and joined Arup as a traffic and transport planner. He gained eight years of experience in carrying traffic impact studies for a wide variety of development within Ireland, using both local and strategic traffic models. He moved back to South Africa in 2009 and continued working for Arup in the Johannesburg office. His work there involved the development of major transport models including a strategic model for Sandton, Waterval and Modderfontein areas. All of these areas are currently some of the fastest growing areas in Gauteng. His work includes roads planning, public transport planning as well as cycle and pedestrian planning and the integration of these modes of transport into the urban fabric of the areas. Simon is contactable at simon.vanjaarsveld@arup.com.

Mr Gerard de Villiers Pr Eng – MSc (Logistics) (Aston, UK), BSc Eng (Civil) (Pretoria), BCom Hons (Transport Economics) (Unisa), BEng Hons (Transportation Engineering) (Pretoria).

Gerard is a professional transportation engineer for more than 36 years who specialises in transport economics, logistics and supply chain management at Arup (Pty) Ltd in Johannesburg since 2011 after four years as Chief Logistics Officer of World Vision International. He started his career as civil engineer at the former Transport Services (Spoornet) in 1979 and subsequently spent time at various consulting engineering companies in transport, transport economics and logistics projects.

He was logistics specialist from 2004 to 2005 at the CSIR where he served as co-leader of the team that developed the first Freight Logistics Strategy for South Africa. More recent projects include pre-feasibility studies for the development of various inland container terminals and logistics hubs as well as the development of a container terminal strategy for Transnet Gauteng terminals, regional corridor development strategy study for the DoT and the automotive supply chain study for Transnet Group Planning.

Gerard has various local and international post-graduate qualifications and teaches part-time in supply chain management at the University of Pretoria as well as the University of Lugano in Switzerland. He was Honorary Professor at the University of Pretoria from 2009 to 2011. He is a Fellow of SAICE, member of various other local and international vocational societies and has presented technical papers at conferences, seminars and vocational societies in South Africa, the US and Australia.

Dr Louis Waldeck has a background in engineering with two degrees in Mechanical Engineering from the University of Pretoria and a PhD in Civil Engineering from the University of the Witwatersrand. He joined the CSIR in 1973 to establish the discipline of building-aerodynamics, leading to research on the dynamic response of slender structures such as masts, towers and chimneys to wind loading. The research involved the development of advanced electronic instrumentation systems, numerical modelling of pneumatic tubing systems and time-series analysis of observations from a network of sensors used to measure wind loading on full-scale structures. He then moved on to the development of expert system technologies for computerising standards and codes of practice (wind and fire) used in the building industry. During the last twenty years of his career, he led various research and development projects associated with management information systems for housing and infrastructure delivery as well as renewable energy.

He is currently responsible for the Urban Dynamics Laboratory aimed at numerically modelling and simulating cities as complex systems using an Urban Simulation Model based on adapted versions of two open-source software packages: UrbanSim and OpenTripPlanner. It simulates the dynamic interaction between government policies, where people live, where they work and how they commute between home and work across various networks over extended periods of time. The adaptations required to achieve success in a developing country like South Africa were recently validated in simulations undertaken for the Cities of Tshwane and Nelson Mandela Bay.

Quintin van Heerden is an Industrial Engineer and Researcher in the Urban and Regional Planning research group of the CSIR Built Environment. His research interests include transport modelling, using UrbanSim as a tool to evaluate current and future spatial planning policies, and data-mining of social media for transport- and urban planning insights. He previously worked as a Transport Modeller and Researcher in commercial vehicle transport modelling in the CSIR Built Environment's Transport & Freight Logistics research group. He has experience in large-scale agent-based transport modelling with a focus on commercial vehicle modelling and the modelling of supply chain stakeholders as agents in an agent-based simulation model, which was also the topic of his Masters in Industrial Engineering. Furthermore, he has a keen interest in mobile technologies, GPS movement analysis, mathematical modelling, operations research, and data-mining. He is a registered candidate engineer at the Engineering Council of South Africa (ECSA) and is working towards professional registration.

Professor Ilse Maria Schoeman holds the position of Associate Professor at the North West University (Potchefstroom Campus) in the Faculty of Natural Sciences in the School for Computer, Statistical and Mathematical Sciences. She specializes in research within the focus of transportation modelling, financial modelling, optimization and transportation planning.

Professor Ilse holds a PhD in Mathematics from the North West University (Potchefstroom Campus) (NWU), South Africa supported by an MSc, BSc (Hons) and BSc degrees all awarded cum laude. She has published internationally in various highly rated journals and serves on various international journal editorial boards. She is also co-author of various chapters in books related to mathematical analysis and modelling. She started to focus on transportation modelling in 2008 and since then contributed to various Urban Transport Conferences of WIT Press. Several Post Graduate students completed their research under her leadership.

She is a member of the Professional Narural Scientists (SACNSP) and Leader of the Sub-programme for Applied Mathematics & Modelling.

Dr James Chakwizira is a Senior Lecturer and Head of Department: Urban and Regional Planning in the School of Environmental Sciences at the University of Venda in South Africa. James previously worked as Head: Roads and Traffic Engineering at the Scientific and Industrial Research and Development Centre (SIRDC). During the same period he deputised the Director – Building Technology Institute. He has also worked as a Senior Researcher and Acting Research Group Leader at the Council for Scientific and Industrial Research (CSIR) initially in the Rural Infrastructure and Services Competency Area and later in the Transport Systems and Operations Competency Area.

Dr James Chakwizira is an experienced spatial and transportation specialist who has consulted on national and regional projects/programmes in Africa, Asia and Europe. He has published, taught at universities in Southern Africa, practiced in both the public and private sector leading multidisciplinary teams and units concentrating on transportation, spatial development, sustainable development, climate change, appropriate technology and sustainable human settlements.

Dr James Chakwizira's passion is to continuously provide transportation thought leadership in the application of new knowledge and technological innovations in order to resolve the challenges of the marginalised and disadvantaged communities and societies inhabiting the built and non-built environments of Africa.

Professor Carel B. Schoeman is a Professor in the Unit for Environmental Sciences and Management at the North West University (Potchefstroom Campus). Carel was involved for 27 years within the local government sector as Urban and Regional Planner, Town Engineer and Director of Planning and Infrastructure. He served in these capacities in several official governmental committees focusing on planning, land use, transportation, resource management and infrastructure. Carel joined the NWU (previously the Potchefstroom University for CHE) in 2002. He has an extensive list of publications in the fields of urban and regional planning, transportation, housing and infrastructure. Some 27 Master's degree and 7 Doctoral research degrees were completed under his leadership. In his career he was involved in various inter-disciplinary project teams. The National Award for Academic Research of the South African Planning Institute (SAPI) was awarded to him in 2016.

Professor Francois Retief is a Professor within the Unit for Environmental Sciences and Management, North-West University. He completed his PhD at the University of Manchester in the United Kingdom on the quality and effectiveness of Strategic Environmental Assessment. He is a professional registered Urban and Regional Planner and has published various high impact papers and chapters in books relating to this topic. Currently he serves on various international journal editorial boards and previously as co-editor of the Journal Impact Assessment and Project Appraisal (IAPA).

Dr Dirk Cilliers is a lecturer in Geography and Environmental Management at the Potchefstroom Campus of the North-West University in South Africa. He holds a Master's degree in Urban and Regional Planning and a PhD in Environmental Sciences. His research interests are focussed around the application of spatial analysis techniques in environmental management and the interface between spatial planning and environmental management. He has been involved in various research projects and has published a number of peer-reviewed papers and book chapters on these topics.

Contents

List of acronyms xiii

Part 1: Transportation modelling

Transportation modelling in practice: connecting basic theory to practice
Werner Heyns & Simon van Jaarsveld 3

Network design and channel strategies
Gerard de Villiers 29

Integrated land-use and transportation modelling in developing countries: using OpenTripPlanner to determine lowest-cost commute trips
Louis Waldeck & Quintin van Heerden 49

The use and application of vector optimisation methods for land use and transportation integration
Ilse Maria Schoeman 69

Part 2: Strategic approach and integration

Transport modelling research agenda and new directions: evidence from a developing country
James Chakwizira 95

Alignment of spatial, transportation and environmental planning: practices underpinning strategic approaches in developing countries
Ilse Maria Schoeman & Carel Benjamin Schoeman 119

Environmental management and land-use management in South Africa: complexities, challenges and opportunities for integrated strategic planning in developing countries
Carel Benjamin Schoeman, Dirk Cilliers & Francois Retief 153

LIST OF ACRONYMS

ABMs	Agent-based models
AHP	Analytic hierarchy process
ANC	African National Congress
APTMS	Advanced public transport management systems
BRT	Bus rapid transport
BSC	Balanced score cards
BV	Binary value
C	Capacity
CGE	Computable general equilibrium
CLARA	Community Land Reform Act
CoG	Centre of gravity
CPD	Continuing Professional Development
C-Planning	Conservation planning
CR	Consistency ratio
CRPs	Commuter Rail Plans
CSIR	Council for Scientific and Industrial Research
CUFM	California Urban Futures Model
DC	Distribution centre
DEAT	Department of Environmental Affairs and Tourism
DFA	Development Facilitation Act
DIUDF	Draft Integrated Urban Development Framework
E	Elasticity coefficient
Eas	Environmental Authorizations
ECSA	Engineering Council of South Africa
EIAs	Environmental Impact Assessments
EIPs	Environmental Implementation Plans
EMF	Environmental Management Frameworks
EMProg	Environmental Management Programme
EMPs	Environmental Management Plans
EMs	Economic Models
eNATIS	National Traffic Information System
EPWP	Extended Public Work Program
4PL	Fourth-party logistics
FOSAD	Forum for South African Directors-General
FTP(s)	Freight Transport Plans
GAM	Goal achievement matrixes
GCR	Gini concentration ratio
GDP	Gross domestic product
GDRT	Gauteng Department of Roads and Transport
GIS	Geographic information system
GITMC	Gauteng Integrated Transport Modelling Centre
GPC	Gauteng Planning Commission
GPMS	Gauteng Pavement Management System
GTFS	General transit feed specification

GTI	Geo terra image
GTS	Gauteng Transport Study
HIA	Heritage impact assessment
HOV	High (3–5 persons) occupant vehicle
IDP(s)	Integrated Development Plan(s)
IEM	Integrated Environmental Management
IEMPs	Integrated Environmental Management Plans
IFM	Integrated Fare Management
IG	Integration of the three spheres of government
IGF	Intergovernmental Forum
ILUMASS	Integrated Land Use Modelling and Transport System Simulation
ILUTE	Integrated Land Use Transportation Environment
IMREL	Integrated Model of Residential and Employment Location
IPTN	Integrated Public Transport Network
IRPT	Integrated Rural Public Transport Planning
IRPTN	Integrated Rapid Public Transport Network
ITE	Institute of Transport Engineers
ITLUP	Integrated Transportation and Land Use Package
ITMP	Integrated Transportation Management Plan
ITPs	Integrated Transport Plans
IUDF	Integrated Urban Development Framework
JMPT	Joint Municipal Planning Tribunal
JV	Joint venture
KPI	Key performance indicators
KZN	KwaZulu-Natal
LDM	Land-Development Management
LDMPs	Land-Development Management Plans
LDOs	Land Development Objectives
LEFTEA	Less Formal Township Establishment Act
LGTA	Local Government Transition Act
LILT	Leeds Integrated Land-Use/Transport model
LM(s)	Local Municipalities
LOS	Level of service
LUM	Land Use Management
LUMS	Land Use Management Systems
LUTI	Land use transportation interaction
LUT	Land use and transport
M&PAs	Monitoring and Performance Assessments
MAUP	Modifiable areal unit problem
MCA	Multi criteria analysis
MCPs	Mine Closure Plans
MDGs	Millennium development goals
METROSIM	Microeconomic land-use and transport model
MINMEC	Intergovernmental Relations Committees of Ministers and Provincial Councils
MLUP	Municipal Land-Use Plans
MPAPs	Monitoring and Performance Assessment Plans
MPRDA	Mineral and Petroleum Resources Development Act
MPT	Municipal Planning Tribunal

MRPs	Municipal Regeneration Plans (Urban Renewal, etc.)
MSDF(s)	Municipal Spatial Development Framework(s)
NAAMSA	National Association of Automobile Manufacturers of South Africa
NATMAP	National Transport Master Plan
NATMAP/SOL	National Transport Master Plan/State of Logistics Survey
NDoT	National Department of Transport
NDP	National Development Plan
NEMA	National Environmental Management Act
NEMBA	National Environmental Management: Biodiversity Act
NEMPAA	National Environmental Management: Protected Areas Act
NFFM	National Freight Flow Model
NFSD	National Framework for Sustainable Development
NHRA	National Heritage Resources Act
NHTS	National household travel survey
NIDS	National income dynamics study
NLSTF	National Land Transport Strategic Framework
NLTTA	National Land Transport Transition Act
NMT	Non-motorized transport
NPC	National Planning Commission
NSDF	National Spatial Development Framework
NSDP	National Spatial Development Plan
NW	North-West
O-D	Origin–Destination
OSM	OpenStreetMap
OTP	OpenTripPlanner
PCC	President's Coordinating Council
PECAS	Production, Exchange and Consumption Allocation System
PLTF(s)	Provincial Land Transport Frameworks
POLIS	Projective Optimization Land Use Information System
PPs	Precinct Plans
PRASA	Passenger Rail Agency of South Africa
PSDF	Provincial Spatial Development Framework (Perspective)
PT	Public Transport
PTPs	Public Transport Plans
R&D	Research and Development
RATPLANS	Rationalization of Public Transportation Service
RCI	Random consistency index
RDPs	Rural Development Plans (Strategies)
RGCI	Row geometric consistency index
RGMM	Row geometric mean method
RPL	Recognition of Prior Learning
RS	Road safety
RSA	Republic of South Africa
RSDF	Regional Spatial Development Framework
RURBAN	Random-Utility URBAN model
S.Dev.Ps	Site Development Plans
SACNASP	South African Council for Natural Scientific Professions
SACPLAN	South African Council for Planners
SALA	Subdivision of Agricultural Land Act

SALGA	South African Local Government Association
SAPIA	South African Petroleum Industry Association
SDF(s)	Spatial Development Frameworks
SDPs	Spatial Development Plans
SE	Spatial interaction gravity-based models
SEA(s)	Strategic Environmental Assessments
SIA	Social impact assessment
SOEs	State-Owned Enterprises
SOV	Single-occupant vehicle
SP	Statutory Planning
SPLUMA	Spatial Planning and Land Use Management Act
SPPs	Statutory Planning Processes
StatsSA	Statistics South Africa
TAZ	Travel analysis zones
3PL	Third-party logistics
TC	Travel cost
TDM	Travel Demand Management
TIAs	Traffic Impact Assessments
TIPs	Transportation Infrastructure Plans
TIS	Transport Impact Studies
TLUMIP	Transport and Land Use Model Integration Program
TMH	Technical methods for highways
TRESIS	Transportation and Environment Strategy Impact Simulator
TSPs	Transportation Sector Plans
TT	Travel time
UN	United Nations
UNEP	United Nations Environmental Programme
UP	University of Pretoria
USA	United States of America
USM	Urban simulation model
V	Volumes

Part 1
Transportation modelling

TRANSPORTATION MODELLING IN PRACTICE: CONNECTING BASIC THEORY TO PRACTICE

WERNER HEYNS[1] & SIMON VAN JAARSVELD[2]
[1]Aurecon Pty Ltd, Pretoria, South Africa
[2]Arup Pty Ltd, Johannesburg, South Africa

ABSTRACT

Transportation planning and modelling can appear to be a complex process when first encountered. Transportation planning is a process that develops information to help make decisions on the future development and management of transportation systems. It involves the determination of the need for new or expanded highways, public transport systems, freight facilities, transportation terminals, their location, their capacity and the management of their demand. Typically, transportation planning involves a forecast of travel patterns and demand 15 to 25 years into the future with an aim to develop a future transportation system that will support spatial growth at that time. Travel demand occurs as a result of thousands of individual travellers making individual decisions on how, where and when to travel. These decisions are affected by many factors such as family situations, characteristics of the person making the trip, and the choices (destination, route and mode) available for the trip. Transportation modelling is used to forecast travel patterns and demand and includes numerous mathematical equations to simulate or represent how people travel.

This chapter explains how the process works, the assumptions made and the steps taken. It also provides an insight into how transportation modelling is applied in practice in demonstrating its role in the inter-relationship between spatial development and transport provision and planning.

Keywords: traffic, modelling, land use, transport integration, four step process, trip generation, trip distribution, mode split, trip assignment, land use, spatial development.

1 INTRODUCTION

The implementation of transport infrastructure and services are derived from transportation planning and modelling outputs. Transportation modelling and simulation as interface in the planning process plays an important role in enabling spatial development.

The models developed by transport planning professionals are simplified representations of reality which can be used to explore the consequences of particular policies, strategies and spatial changes. The models forecast transport demand based on expected spatial development, population and economic growth, which translates in the creation of additional trips on the road network requiring expansion or the provision of infrastructure for a variety of modes.

The travel forecasting process is at the heart of transportation planning (Goulias [1]). The process is used to estimate the number of trips that will be made on a transportation systems alternative at some future date. It involves a series of mathematical models that attempt to simulate human behaviour while traveling (Tri-County Regional Planning Commission [2]). The models include a sequence of steps that answer a series of questions about traveller decisions. Attempts are made to simulate all choices that travellers make in response to a given system of highways, public transport and policies. Many assumptions need to be made about how people make decisions, the factors they consider and how they react to a particular transportation alternative.

This chapter is broadly divided into two sections. The first 8 sub-sections provide basic theory and an overview of transportation modelling processes in general, considering the what, why, who and how of modelling and actual models used to forecast future travel.

The second component of the chapter defines the practical application of modelling software, how it relates land use changes to infrastructure provision, the type of problems they seek to address, the decisions planning professionals have to make in deciding about the appropriate model in terms of applicability, costs and data requirements. Fig. 1 illustrates the components covered in this Chapter.

2 LAND-USE AND TRANSPORT INTEGRATION

The control and planning of land use is to a large extent the key to the control of both the demand for transport and its impact upon the environment. Although the inter-relationship between land use and transport is complex, it is well established. An inherent interaction or feedback process takes place between land use and transport, where the one influences and depends on the other. (American Association of State Highway and Transportation Officials [3], Toth [4]). Wegener [5] refers to the interaction as the land use transport feedback cycle, illustrated by Fig. 2.

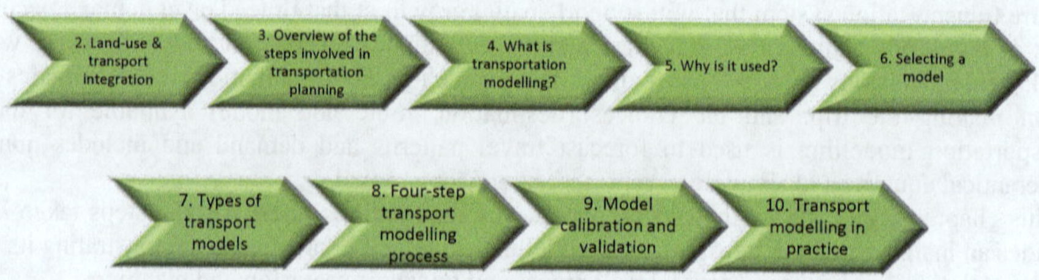

Figure 1: Components of transportation modelling examined.

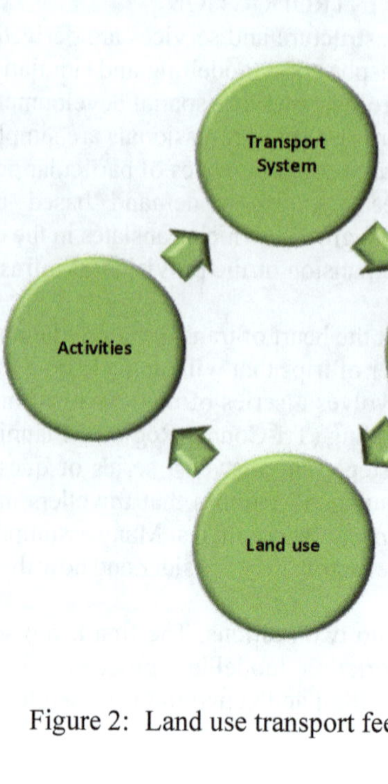

Figure 2: Land use transport feedback cycle.

Starting at the bottom of the cycle (Land Use), the locations of population and employment determine the origins and destinations of most trips in travel models (Activities). Modelling the transportation system allows calculating accessibilities, which describe how well accessible a zone is to all other zones. Accessibility shapes land use, as both households and businesses search for locations that are – among other location factors – accessible. Because of this interaction, it is important for transport and land use planning professionals to understand the likely inter-relationships and possible negative side effects when planning is not executed within an appropriate framework. For example, a land use policy aimed at encouraging urban sprawl, does little to promote shorter work-home trips, reduced travel cost and the financial requirements of operating and funding public transport. Counter intuitive and counterproductive land use and transport planning has serious long lasting consequences. In South African cities, the legacy of Apartheid is a good example demonstrating the transport inefficiencies resulting from inappropriate land policies, including:

- Misguided investment in transportation systems that prioritize high speed mobility over local accessibility (This places all other modes of travel at a disadvantage);
- Urban sprawl;
- Under provision of Non-Motorised Transport Infrastructure;
- Commuter public transport services focused on operating mainly during the morning and afternoon peak hours, with limited inter peak services;
- High travel cost (private and public transport) – mobility and accessibility is expensive and has a high economic and social cost;
- Rural communities often isolated and inaccessible, contributing to the inability to access opportunities and jobs resulting in perpetuated poverty.

Transportation modelling is a strong and helpful tool in the transport planner's tool kit that supports decisions making. It will not eradicate poor decision making or controversial land use policies, but some major advantages of using models in transport planning in supporting integration is that:

- it estimates likely land use transport interactions more quickly and at lower cost and risk than would be through implementation and monitoring or trial by error;
- it helps planners optimise the balance between land use provision and appropriate transport infrastructure and relevant modes to enable sustained economic growth;
- it defines the appropriate level of infrastructure required to support land use and spatial development;
- it defines the appropriate modes of transport to connect land use;
- it weighs up benefits/dis-benefits of various development scenarios; and
- it informs investment decisions by providing input into financial, economic, environmental and other relevant forms of appraisal.

3 OVERVIEW OF THE STEPS INVOLVED IN TRANSPORTATION PLANNING

Before exploring what transport modelling is, it is necessary to understand the broad steps involved in transportation planning as to understand how transportation modelling fits into the overall planning process.

Transportation planning goes through a basic sequence of steps. Several can take place simultaneously and it is not unusual to repeat some of the steps (United States Department of Transportation Federal Highway Authority of Transportation [6], [7]). Fig. 3 illustrates the broad overall planning process.

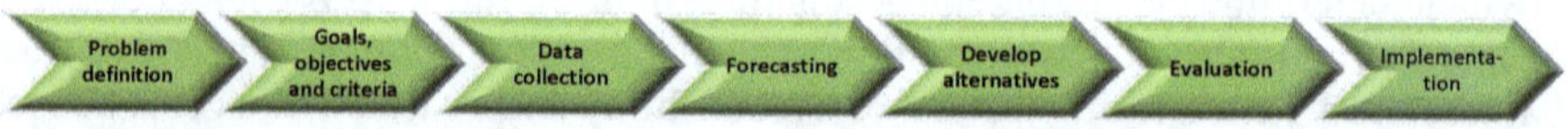

Figure 3: Basic steps in the transportation planning process.

- **Problem definition:** What is the key transportation, economic and land use issues and problems facing the community? This step may also involve definition of the size of an area to be studied, determination of the scope of the study and the establishment of a steering committee structure or forum to oversee the planning process.
- **Define goals, objectives and criteria:** A consensus should be developed by elected officials and citizens about the future of the community and its transportation system. Goals are developed for the quality of transportation service, environmental impacts and costs and are typically in conflict. A good planning effort will identify the trade-offs between these factors among alternatives in a clear, concise way to help make decisions.
- **Data collection:** Data must be compiled about the present status of the transportation system and its use. This could include traffic data, public transport ridership statistics, census information and interviews of households about their travel patterns. Data are also gathered on land use, development trends, environmental factors, and financial resources. This will help in problem definition and in developing methods to forecast future travel patterns. Good data are essential to the planning process. The statement 'garbage in/garbage out' applies in transportation planning. Without good data, the results of the planning process have little real meaning and can lead to wrong project selection and an inappropriate developmental direction for the region.
- **Forecasts:** Data from existing travel is used to make forecasts of future travel using travel demand models. This requires forecasts of future population, land use and economic conditions as well as understanding of how people make travel choices. Forecasting requires large amounts of data and is carried out under many assumptions which is based on benchmarking, sensitivity checks and follow up surveys to identify trends.
- **Develop alternatives:** Forecasts are used to determine the performance of alternative future land use and transportation systems. Alternatives normally include different land use and transportation patterns with different mixtures of highway and transit services and facilities. Since land use affects travel and travel affects land use, both must be considered.
- **Evaluation:** Results of forecasts are used to compare the performance of alternatives in meeting goals, objectives and criteria. This information may be extensively discussed by interested citizens, elected officials, different government agencies and the private sector. Ultimately decisions are made by appropriate elected or appointed officials and groups.
- **Implementation of plans:** Once decisions are made, plans should be further developed and refined for implementation. This may include more detailed analysis for design and evaluation following a similar process as above.

4 WHAT IS TRANSPORTATION MODELLING?

Whilst many definitions exist defining transportation modelling it is primarily a mathematical tool using computer software to represent an actual transport system (the real world) to forecast travel patterns and flows between origins and destinations in geographic space by different modes (Victoria Transport Policy Institute [8]). The model output informs decisions about the type of infrastructure and transport modes to be implemented or operated in support of changing land use and travel patterns.

For example, when a new road is to be built, a transport model is usually developed to see how much traffic the road will carry, how wide it should be, where to include road intersections and defining the appropriate type of intersection control etc. After all, if you are able to save the cost of a slip road or the implementation of an additional lane on a highway the cost of developing and using the model is only a fraction. Because the investment requirement of transport infrastructure is very high and most often a burden on the fiscus clear advice is required to government or developers to provide value for money solutions. Transport models also sometimes cover public transport, walking and cycling and at the other end of the scale they cover air, sea and freight transport. These models may also include fares where the higher the fare the less people will use the infrastructure facility or mode of transport. This gives rise to the notion of elasticity in demand (Litman [9]), which is a relationship between a fare and the number of people who use a transport service or infrastructure.

5 APPLICATION OF TRANSPORTATION MODELS

Transportation models informs decisions about the type of infrastructure and transport modes to be implemented or operated in support of changing land use and travel patterns (Chandrasekaran [10]). Modelling helps to gain insight into complex situations by understanding simpler situations resembling them (Florian [11]). Professionals in transport and land use development also apply transportation models for some of the following reasons:

- Forecasting traffic flow;
- Assessing the outcomes of alternative land uses, infrastructure network and policy alternatives:
  - For smaller areas – models are good for considering alternative designs (e.g. different sized roundabouts, intersection layouts etc.);
  - For larger areas – to answer strategic questions such as whether or not to build new roads or implement urban pricing;
- Corridor studies;
- Supporting development control policy testing;
- Land-use transport interaction;
- Understanding travel behaviour;
- Establishing how to reduce traffic congestion;
- Establishing how to reduce fuel consumption;
- Establishing how to improve air quality;
- Optimizing the use of resources in building or maintaining transport systems; and
- Providing a platform for testing competing ideas and use in negotiations and decision-making.

Transportation models are used in a sequence of steps to answer a series of questions about future travel patterns that supports forecasting demand and ultimately planning decisions.

The following questions follow a logical sequence and inform the steps in the modelling process:

1. What will our community/city/region look like in the future?
   a. How many people will live there? (population forecasts)
   b. What will they be doing? (economic forecasts)
   c. Where will activities take place? (land use)
2. What are the travel patterns in the future?
   a. How many trips will be made? (trip generation)
   b. Where will the trips be? (trip distribution)
   c. What modes will be used? (mode split)
   d. What routes will be used? (traffic assignment)
   e. What will be the effects of this travel? (impact analysis)

Ultimately transportation models should help the decision maker save money, define the appropriate level of transport investment required in support of spatial changes and implement value for money transport projects, systems and operations. Most modelling work is concerned with appraisal, in the broadest sense of the word, of alternative options or designs.

6 SELECTING A MODEL

When a range of models is available, the choice is influenced by the level of complexity of the problem being sought to be addressed. Models range in complexity from simple equations representing relationships which can be calculated on the back of an envelope, on a calculator in a spreadsheet or using powerful off the shelf (commercial) software programmes. The latter involving a large volume of coding to perform sophisticated mathematical functions or detailed simulations.

When selecting a model, the user should begin by defining the requirements of the exercise to be undertaken and the resources required. The requirements that should be defined include:

- What input variables are required, and to what level of detail?
- What output variables are required, at what level of detail and in what form (e.g. on screen results or graphics and results to "come alive" where the movement of cars are observed on screen)?
- Geographical coverage (will a single intersection be simulated, a number of intersections or an entire suburb?)
- Constraints of time and money;
- Is it possible to buy new commercially available software or should new software or processes be developed? and
- Determine whether new data can be collected.

The objective of defining the requirements for analysis to determine up front the most appropriate model for the task at hand, to save money and time and derive the right answer.

7 TYPES OF TRANSPORTATION MODELS

A variety of types of models is available to the analyst. The different traffic models each have its own pros and cons. The type of network under study and the traffic characteristics strictly influence the choice of the traffic model used for analysis. There is no one single model that can be used effectively for modelling traffic in all kinds of networks (Chandrasekaran [10]). For simplicity, two main groups exist:

- Abstract mathematic models; and
- Simulation models.

The review provided below is not exhaustive and highlights the most common models within each group.

7.1 Abstract mathematic models

A number of abstract mathematical models exist including the following:

- **Simple formulae:** An equation of the general form $y = f(x)$ can be used to represent an empirically observed relationship between one variable *(x)*, which can be taken as given, and another *(y)* which is to be predicted.
- **Time series models:** Many aspects of transport demand vary over time. For example, car ownership and total vehicle- kilometres have been rising almost continually over the past few decades, except for blips coinciding with fuel crises or economic recession. By plotting the relevant data over time, it becomes possible to deduce and underlying trend and or relationship. A prediction of future levels of travel demand can be made by extrapolating the trend into the future. Trend based forecasting is very popular because of its simplicity and because it is based on past evidence (Brockwell et al. [12]).
- **Averaging and smoothing models:** When no apparent trend data is available to support demand forecasting, simple arithmetic can be used to calculate an average or an exponentially smoothed prediction. Average forecasts are simply but very crude and may give a distorted view, exponential smoothing seek to overcome the crudeness, by giving weights to the most recent and trustworthy data.
- **Regression analysis:** This is the process of identifying the mathematical function (relationship) which best fits the observed data. The most common applications are in the prediction of car ownership, trip ends (numbers of origins and destinations at a given location) and trip volumes (in an area between a given pair of zones or along a link) (Jha et al. [13]). It is also used to predict waiting times at intersections, numbers of accidents or levels of pollution.
- **Matrix estimation models:** The cells of a demand matrix (known as an origin – destination matrix) indicate the number of trips between each origin and destination pair, and the row and column totals indicate the total number of origins and destinations respectively in each zone. The matrix provides the overall picture of travel demand in a study area and is a key component in transport analyses.
- **Elasticity models:** When short term planning is undertaken, such that some influences can be considered constant, it is possible to specify a model which is solely concerned with the sensitivity of the forecast to key policy variables – known as elasticity modelling The elasticity of demand with respect to a certain variable is defined as the rate of change of demand with respect to that variable, normalized by the current levels of demand and the variable in question (Meyer and Miller [14]). Elasticity is thus a measure of sensitivity to change in system conditions (He and Zhao [15]). Its general form is:

  $$Y_t = Y_{t-1}\{1+E(X_t - X_{t-1})/X_{t-1}\},$$

  where, Y_t = quantity demand in year t, X_t = value of a supply variable (e.g. price or journey time) in year t, and E is the elasticity coefficient for Y with respect to X.

The elasticity coefficient should be calibrated on past data containing evidence of the marginal effect on *Y* of a marginal change in *X*.

- **Demand allocation – modelling the choice between alternatives:** Trips are made as a result of choices made by travellers; choices between alternative modes of travel, alternative times of travel and alternative routes. Some of these choices are heavily constrained (e.g. if the traveller has no car available or if a public transport timetable provides only one service per day) while others may be made from a very extensive choice set. Different groups in the population have different constraints and those who have a very restricted choice set may be described as captives of a particular option. Even if they are not captive to one option some groups may have a strong preference for one option over the others; for example, high income travellers are likely to have a high value of time which causes them to prefer a fast and expensive option over a slower and cheaper one.

  There is a wide and growing range of models which seek to predict travellers' choices among available options. They vary in sophistication from models which simply aim to reproduce an observed pattern to others which seek to replicate the underlying choice processes. An important concept underlying many of these models is market segmentation whereby separate forecasts are made for subgroups within the total population who, by virtue of their particular characteristics of car ownership, income, journey purpose etc., may be expected to respond to the choices available differently from other subgroups. Having made a forecast, for each subgroup, of the proportion of group members selecting each option, the overall forecast is simply achieved by weighting the proportions by the expected size of each subgroup in the forecast year's population.

7.2 Simulation models

Simulation models differ from abstract mathematical models in that they attempt to represent the dynamic evolution of some aspect of the transport system through an explicit representation of the behaviour of actors within it. Therefore, at a detailed scale, the performance of an intersection might be simulated by representing individual cars passing through it and the performance of a bus route might be simulated by representing the boarding and alighting of individual passengers at each stop. At a less detailed level, the development of a bus network might be predicted as the result of competition between different operating companies.

Another major difference between simulation and abstract modelling is that simulation modelling allows the analyst to develop a logic model of a very complex system by putting together a number of components which are themselves fairly well understood. It therefore becomes possible to develop a model of a complex system without having recourse to overarching theories about system dynamics or concepts such as equilibrium. When simulation is very detailed (zooming into fine detail analysis), it is sometimes referred to as micro-simulation, but the term is sometimes used interchangeably. When applied in traffic engineering the units are vehicles. When applied in travel behaviour the units are persons and households (Goulias [1]).

With cheaper, readily available commercial software programmes and ease of use and understanding opposed to abstract mathematical models, simulation modelling is undoubtedly the fastest growing area in modelling and are more often applied in solving transport and traffic problems.

Boxill and Yu [16] have classified simulation models into three categories:

- Macroscopic;
- Mesoscopic models; or
- Microscopic models;

Simulation models can also be classified according to the nature of the network that they can be applied to, e.g. signalized networks, highway networks, and integrated networks or for specific purposes (Variable road pricing, etc.). Tables 1 to 3 illustrate a range of simulation modelling software programmes according to the classification by Boxill and Yu [16].

Table 1: Macroscopic models.

Model	Classification
ARTWORK	Highway networks
AUTOS	Highway networks
CORFLO	Integrated Networks
FREFLO	Highway networks
FREQ	Highway networks
KRONOS	Unknown
METACOR	Unknown
METANET	Highway networks
NETFLO	Signalized networks
NETFLO 2	Signalized networks
NETVACI	Unknown
PASSER-II	Signalized networks
PASSER-IV	Signalized networks
TRANSYT-7F	Signalized networks
TRANSYT/10	Signalized networks
TEXAS	Signalized networks

Source: Boxill and Yu [16].

Table 2: Mesoscopic models.

Model	Classification
DYNAMIT	Integrated Networks
DYNEMO	Integrated Networks
DYNASMART	Integrated Networks

Source: Boxill and Yu [16].

Table 3: Microscopic models.

Model	Classification	Model	Classification
AIMSUN	Integrated Networks	PLANSIM-T	Highway networks
ANATOLL	Tolls	ROADSIM	Rural road network
ARCADY	Signalized networks	SATURN	Integrated Networks
AUTOBAHN	Highway networks	SHIVA	Other
AVENUE	Signalized networks	SIGSIM	Signalized networks
CARSIM	Signalized networks	SIMCO2	Highway networks
CASIMIR	Signalized networks	SIMDAC	Highway networks
CONTRAM	Integrated Networks	SIMNET	Highway networks
CORSIM	Integrated Networks	SIMTRAFFIC	Signalized networks
DRACULA	Integrated Networks	SISTM	Highway networks
FLEXYT II	Signalized networks	SITRA-B+	Integrated Networks
FOSIM	Highway networks	SIDRA	Integrated Networks
FREEVU	Highway networks	SITRAS	Integrated Networks
FRESIM	Highway networks	SMARTAHS	Unknown
HIPERTRANS	Signalized networks	SMARTPATH	Highway networks
HUTSIM	Signalized networks	SOUND	Integrated Networks
ICARUS	Integrated Networks	SPEACS	Highway networks
INTEGRATION	Integrated Networks	STEER	Highway networks
INTRAS	Signalized networks	STREETSIM	Signalized networks
JAM	No Available Data	TEXAS	Signalized networks
MELROSE	Integrated Networks	TEXSIM	Signalized networks
METROPOLIS	Highway networks	TRAFFICQ	Signalized networks
MICROSIM	Signalized networks	TRANSIMS	Highway networks
MICSTRAN	Signalized networks	TRGMSM	LRT
MIMIC	Safety	TRITRAM	Highway networks
MITSIM	Integrated Networks	THOREAU	Integrated Networks
MIXIC	Highway networks	UTSS	Integrated Networks
NEMIS	Integrated Networks	VEDENS	Signalized networks
NETSIM	Integrated Networks	VISSIM	Integrated Networks
OLSIM	Highway networks	VISSUM	Integrated Networks
PADSIM	Highway networks	WATSIM	Integrated Networks
PARAMICS	Integrated Networks	WEAVSIM	Integrated Networks
PELOPS	Integrated Networks		

Source: Boxill and Yu [16].

8 FOUR-STEP TRANSPORT MODELLING PROCESS

Travel demand modelling was first developed in the late 1950s as a means to assist in highway planning. As the need to look at other problems and issues arose, the transportation modelling process was modified subsequently to add additional techniques to deal with these problems.

As a starting point one can ask, how is the city or region represented for computer analysis? Travel simulation require that an urban area or region be represented as a series of small geographic areas referred to as travel analysis zones (TAZs). Fig. 4 illustrates the travel analysis zones and links concept. Zones are characterized by their population, employment and other factors and are the places where trip making decisions are made (trip producers) and the trip need is met (trip attractors). Trip making is assumed to begin at the centre of activity in a zone (zone centroid).

Trips that are very short, that begin and end in a single zone (intrazonal trips) are usually not directly included in the forecasts. This limits the analysis of pedestrian and bicycle trips in the process. Zones can be as small as a single block but typically are 1/4 to one mile square in area. A planning study can easily use 500–2000 zones. A large number of zones will increase the accuracy of the forecasts but require more data and computer processing time. Zones tend to be small in areas of high population and larger in areas of low density and population. Internal zones are those within the study area while external zones are those outside of the study area. The study area should be large enough so that nearly all (over 90%) of the trips begin and end within the study area. The trips that do not begin or end within the study area but travel through it is considered to be through traffic.

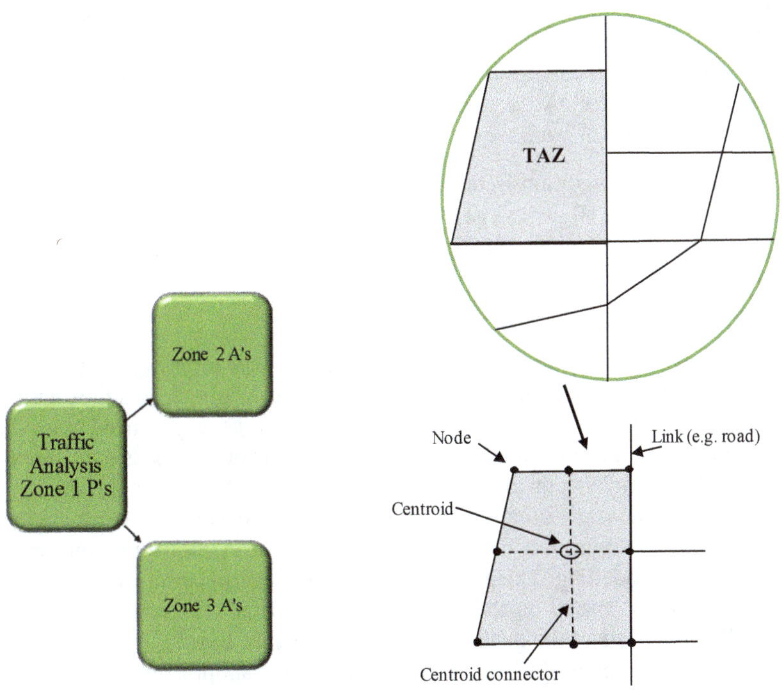

Figure 4: Travel analysis zones.

The highway system and transit systems are represented as networks for computer analysis. Networks consist of links to represent segments of highways or public transport lines and nodes to represent intersections and other points on the network. Data for links includes travel times on the link, average speeds, capacity, and direction. Node data is more limited to information on which links connect to the node and the location of the node (coordinates). Node data could also include data on intersections to help calculate delay encountered at intersections.

The travel simulation process follows trips as they begin at a trip generation zone, move through a network of links and nodes and end at a trip attracting zone. The simulation process is known as the four-step process and includes (i) trip generation, (ii) trip distribution, (iii) mode split and (iv) traffic assignment (Tri-County Regional Planning Commission [2], Victoria Transport Policy Institute [8], Florian [11]).

8.1 Trip generation

The first step in travel forecasting is trip generation. In this step information from land use, population and economic forecasts are used to estimate how many trips will be made to and from each zone. This is carried out separately by trip purpose. Some of the trip purposes that could be used are: home based work trips (work trips that begin or end at home), home based shopping trips, home based other trips, school trips, non-home based trips (trips that neither begin or end at home), truck trips and taxi trips. Trips are calculated based on the characteristics of the zones. Trip productions are based on household characteristics such as the number of people in the household and the number of cars available. For example, a household with four people and two cars may be assumed to produce 3.00 work trips per day. Trip attractions are typically based on the level of employment in a zone. For example, a zone could be assumed to attract 1.32 home based work trips for every person employed in that zone. Trip generation uses trip rates that are averages for large segments of the study area. In South Africa, the following sources are used providing empirical estimates of trip generation per trip purpose or land use type:

- South African Trip Generation Rates Manual [17];
- Technical Methods for Highways (TMH) 17, Trip Data Manual [18]; and
- Institute of Transport Engineers (ITE), Trip Generation Manual [19].

8.2 Trip distribution

Trip generation only finds the number of trips that begin or end at a particular zone. The process of trip distribution links the trip ends to form an origin–destination pattern. Trip distribution is used to represent the process of destination choice, e.g. "I need to go shopping but where should I go to meet my shopping needs?" Trip distribution leads to a large increase in the amount of data which needs to be dealt with. Origin–destination tables are very large. For example, a 1200 zone study area would have 1,440,000 possible trip combinations in its O–D table for each trip purpose. The most commonly used procedure for predict trip distribution is gravity modelling. The gravity model takes the trips produced at one zone and distributes to other zones based on the size of the other zones (as measured by their trip attractions) and on the basis of the distance to other zones (see Fig. 5).

A zone with a large number of trip attractions (say a large shopping centre) will receive a greater number of distributed trips than one with a small trip attraction (a small shopping centre).

	To			
From	1	2	3	4
1				
2				
3				
4				

Intra-zonal trips are not assigned to the road network that connects inter-zonal traffic

Separate tables are used for each trip purpose

Figure 5: Origin–destination table.

Distance to possible destinations is the other factor used in the gravity model. The number of trips to a given destination decreases with the distance to the destination (it is inversely proportional). For example, you would expect more trips to a nearby shopping centre than one further away. The distance effect is found through a calibration process which gives travel times to destinations from the model similar to that found from field data. "Distance" can be measured in several ways. The simplest way this is carried out is to use car travel times between zones is as the measurement of distance.

Other ways might be to use a combination of car travel time and costs such as tolls as the measurement of distance. Still another way is to use a combination of public transport and car times and costs (composite cost). This method involves using a percentage of the car time and cost and a percentage of the public transport time/cost. Because of calculation procedures, the model must be used/run a number of times (iterated) in order to balance the trip numbers to match the initial values.

8.3 Mode split

Mode split or mode choice is one of the most critical parts of the demand modelling process. It is the step where trips between a given origin and destination are split into trips using public transport, trips by car pool or as car passengers and trips by car drivers – see Fig. 6.

Calculations are conducted that compare the attractiveness of travel by different modes to determine their relative usage.

All proposals to improve public transport or to change the ease of using the cars are passed through the mode split/car occupancy process as part of their assessment and evaluation. It is important to understand what factors are used and how the process is conducted in order to plan, design and implement new systems of transportation.

Mode split is carried out by a comparison of the "disutility" of travel between two points for the different modes that are available. Disutility is a term used to represent a combination of the travel time, cost and convenience of a mode between an origin and a destination.

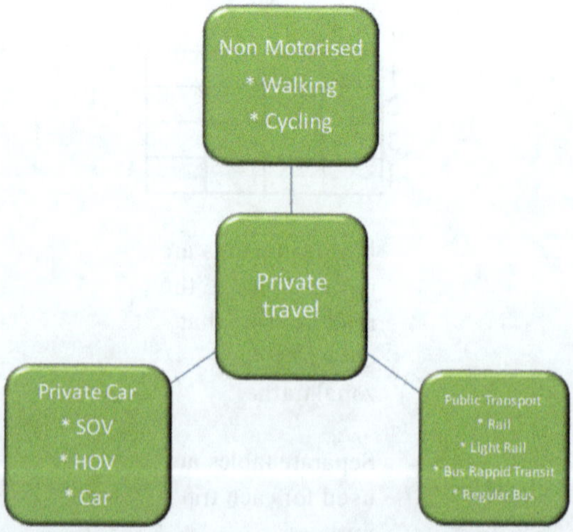

Figure 6: Mode split.

The disutility equations do not normally recognize differences within travel modes. A bus system and a rail system with the same time and cost characteristics will have the same disutility values. It is possible that mode bias factors could be different for different technologies, to represent a preference for light rail over bus, but this must be specifically included.

Once disutilities are known for the various choices between an origin and a destination, the trips are split among various modes based on the relative differences between disutilities. A large advantage will mean a high percentage for that mode. Splits are calculated to match splits found from actual traveller data. Sometimes a fixed percentage is used for the minimum transit use (percent captive users) to represent travellers who have no transport available or are unable to use a car for their trip.

Mode split and car occupancy analysis can be two separate steps or can be combined into a single step, depending on how a forecasting process is set up. In the simplest application, a highway/public transport split is made first which is followed by a split of car trips into car driver and car passenger trips. More complex analysis splits trips into multiple categories (single occupant auto (SOV), two-person car pool, 3–5-person car pool (HOV), van pool, local bus, express bus, etc.). Car occupancy analysis is often a highly simplified process which uses fixed car occupancy rates for a given trip purpose or for given household size and car ownership categories. This means that the forecasts of car-pooling are insensitive to changes in the cost of travel, the cost of parking, the presence of special programs to promote car-pooling, etc.

8.4 Traffic assignment

Once trips have been split into highway and public transport trips, the specific path that they use to travel from their origin to their destination must be found (Chiu et al. [20]). These trips are then assigned to that path in the step referred to as traffic assignment. Traffic assignment is the most time consuming and data intensive step in the process and is performed differently for highway trips and public transport trips. The process first involves the calculation of the shortest path from each origin to all destinations (usually the minimum time path is used). Trips for each O–D pair are then assigned to the links in the minimum path and the trips are added up for each link. The assigned trip volume is then compared to the capacity of the link to see if it is congested. If a link is congested the speed on the link needs to be reduced to result in a larger travel time on that link. When speeds and travel times are changed, the shortest path may change. Hence the whole process must be repeated many times (iterated) until there is an equilibrium between travel demand and travel supply.

Trips on congested links will be shifted to uncongested links until this equilibrium, condition occurs. There are a variety of ways in which the calculations are carried out to reach network equilibrium, in order to keep the computer time to a minimum. One way to get a feel for the accuracy of the models is to look at the resulting speeds on the network. These should be realistic after equilibrium.

Public transport trip assignment is carried out in a similar way except that public transport headways are adjusted rather than travel times. Public transport headways (minutes between consecutive vehicles) affect the capacity of the public transport route. Low headways mean that there is more frequent service and a greater number of vehicles. Public transport supply and demand are also recalculated to reach equilibrium between supply and demand.

It is important to understand the concept of equilibrium. If a highway or public transport route is congested during the peak hour, its excess trips will be shifted to alternative routes. If the alternative routes are also congested the final results will show congestion over a wide part of the network. In the real world, this congestion will eventually dissipate over time.

Another important step in assignment is the time of day analysis. Daily trip patterns need to be converted to peak time period traffic. A key assumption needed is the portion of daily travel that occurs during the peak period. This is normally used as a constant and conventional travel models have very limited capability to describe how travellers will shift their trips to less congested times of the day.

Some of the assumptions in traffic assignment are as follows:

- **Delay occurs on links:** Most traffic assignment procedures assume that delay occurs on the links rather than at intersections.
- **Travel only occurs on the network:** It is assumed that all trips begin and end at a single point in a zone (the centroids) and occurs only on the links included in the network. Not all roads streets are included in the network nor all possible trip beginning and end points included. The zone/network system is a simplification of reality.
- **Capacities are simplified:** To determine the capacity of roads and public transport systems requires a complex process of calculations that consider many factors. In most travel forecasts this is greatly simplified.
- **Time of day variations:** Traffic varies considerably throughout the day and during the week. The travel demand forecasts are made on a daily basis for a typical weekday and then converted to peak hour conditions. Daily trips are multiplied by a "hour adjustment factor", for example 10%, to convert them to peak hour trips. The number assumed for this factor is very critical. A small variation, say plus or minus one percent, will make a large difference in the level of congestion that would be forecast on a network. Most models are unable to represent how travellers often cope with congestion by changing the time they make their trips.
- **Emphasis on peak hour travel:** As described above, forecasts are carried out for the peak hour. A forecast for the peak hour of the day does not provide any information on what is happening the other 23 hours of the day. The duration of congestion beyond the

peak hour is not determined. In addition, travel forecasts are made for an "average or typical weekday". Variations in travel by time of year or day of the week are usually not considered.

9 MODEL CALIBRATION AND VALIDATION

Transportation models are aimed at reflecting an open system – the real world populated by people who are responding to influences that are constantly changing; do not always make rational decisions, and whose responses to influences affecting travel are not always the same. Since transportation models (and travel modellers) cannot be omniscient, there will always be missed information and abstractions resulting in less than perfect models.

Therefore, once the model is estimated and or set up using the four-step process, small adjustments are made to it until the model accurately replicates observed patterns and behaviours – referred to as calibration. Then the model is validated to a different set of observed data (for example, traffic counts) than was used for model estimation or calibration (United States Department of Transportation Federal Highway Administration [21], Wegmann and Everett [22]).

- **Model calibration** is the adjustment of constants and other model parameters in estimated or asserted models in an effort to make the models replicate observed data for a base year or otherwise produce more reasonable results;
- **Model validation** is the application of the calibrated models and comparison of the results against observed data; ideally, the observed data are not the same data used for model estimation or calibration.

Model calibration and validation generally occur in an iterative fashion. Model validation may reveal the need to return to the model estimation or model calibration steps. The application of the model using future year conditions requires that the model forecasts are reasonable and consistent with expectations and also might reveal a need to return to the model estimation or calibration step.

10 TRANSPORTATION MODELLING IN PRACTICE

Once a model has been developed, calibrated and validated it is ready for application, assessing policy decisions, testing the outcome of different scenarios or alternatives. This section outlines a number of examples where transportation modelling has been used to explore and solve "real life" problems – joining theory and practice. The examples demonstrate the use of models on a variety of levels. From the local level site specific at the one end to a regional level at the other end. To protect client confidentiality, street names and features that may identify buildings or land marks have been removed from the figures and images in this section.

The examples provided are presented in the form of seven case studies outlining the type of problem that was encountered, the type of modelling software used to address the problem and why and the outcome and or benefits of the modelling exercise.

Case study 1

The first case study relates to the local level – a site specific development, where a practical problem is considered. That is: "what is the sizing/vehicle capacity requirement of the site's vehicle entrance(s) to accommodate the forecasted vehicle demand?" When determining the number of vehicle entrances, there are two factors that determine ingress and egress capacity:

(i) the number of lanes provided; and
(ii) the distance between the car park barrier/gates and the public road (also referred to as vehicle stacking distance).

If there are too little gate/lane capacity or the vehicle stacking distance is too short there is a risk that vehicles will spill over into the public road or stack on the public road causing congestion, road safety issues and traffic on the public road being negatively impacted (see Fig. 7 as a general illustration).

This is a very common problem and question when faced with new site developments. The success of the development depends on having sufficient access and the accesses operating optimally. If it does not the land owner/shop keeper may lose business as shoppers may consider going elsewhere if it is simply too difficult to access.

However, there is a very simple solution to solving the problem in isolation. A vehicle queuing model, determining the probability of vehicles that may stack, based on expected vehicle volumes and density, gate service levels and number of gates provided can be used as a suitable modelling tool to determine the optimum access requirements. It is inexpensive and data input requirements are limited.

Case study 2

The problem discussed in case study 1 can be expanded to a more complex problem where for example a development site has multiple accesses over which traffic is distributed into a large office building (as example). The question that arises is: "how do we design the car park optimally given a number of accesses and the inter-operability between the levels of structured or basement parking.'' This problem is compounded when the parking facility is a multi-storey (see Fig. 8) or underground car park with many parking levels accommodating a few thousands parking bays. The attraction of vehicles to each access is based on an 'attraction' factor, determined by the perceived access to parking bays. Whilst the distribution of vehicles around the site was determined during the trip distribution step in the four-step model, it did not address how the interaction of vehicles will occur once within the building and drivers are hunting parking spaces.

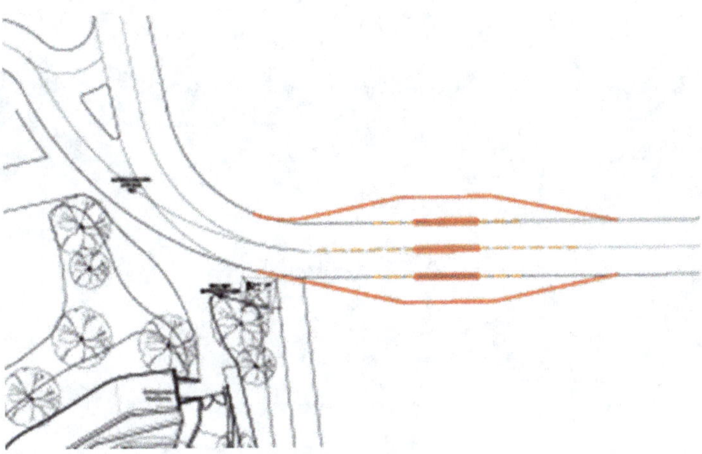

Figure 7: Determining vehicle access lane capacity.

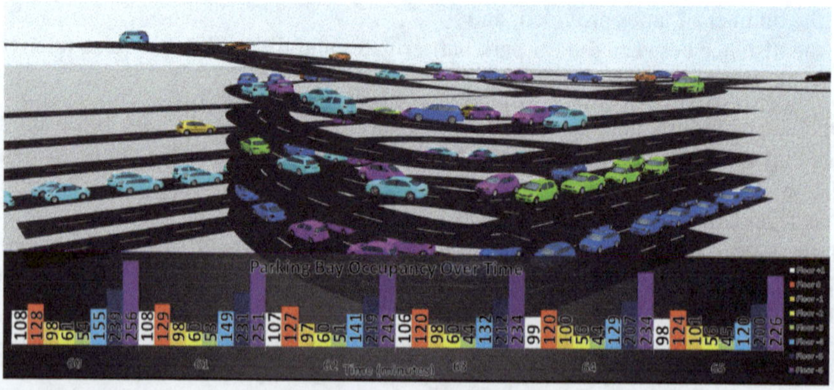

Figure 8: Structured parking.

Although using a spreadsheet and queuing analysis model in isolation will provide some clues to solve the problem it is incapable of solving the problem entirely due to the dynamic nature of the problem. Why? Because simple queuing analysis models consider one access problem at a time and not the inter relationship between different accesses along the same street or from different streets to the same site or building. The problem is further exacerbated where traffic streams at different accesses meet within the basement or structure parking as a result of random internal basement route choice.

To deal with this problem and the inter-related driver choices and behaviour dynamics, micro-simulation modelling with vehicle assignment functionality has proven to be a suitable modelling tool. Fig. 9 illustrates a screen shot of a micro simulation model being used to determine access requirements to a large-scale development.

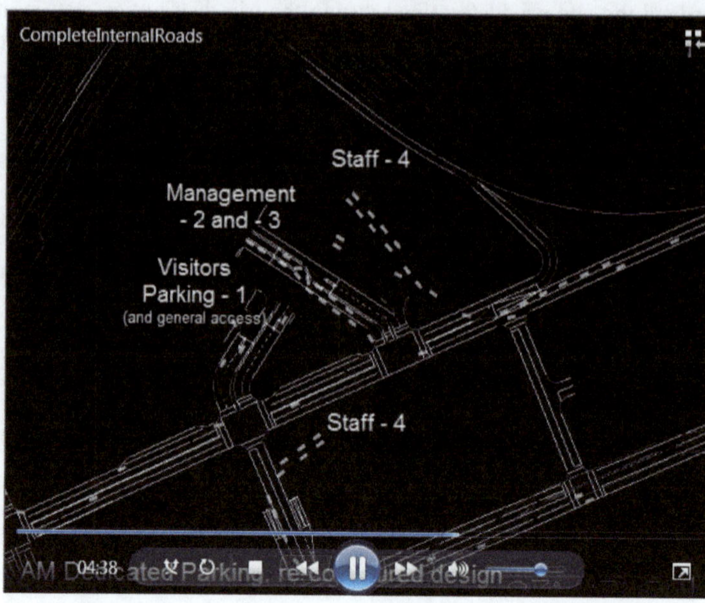

Figure 9: Micro-simulation modelling and parking design.

One of the benefits of micro-simulation modelling is the visual presentation of the modelling results. Non-traffic engineers/transport planners are able to observe the 'logic' of how traffic patterns occur in and around the site/building and therefore reduce scepticism about whether the solution may be effective or not. The model is often used to demonstrate to clients and local authorities that a specific access strategy works optimally or that some other solution may be required to deal effectively with forecasted traffic volumes.

Case study 3

The previous two case studies examined access problems at the local or site specific level. However, traffic engineers and transport planners often have to assess the operation of intersections/interchanges on public roads, which are off site. Typically, such assessments are necessary as part of Traffic Impact Assessment studies or Environmental Impact Assessments. The challenges/problems faced here include some of the following:

- What is the most suitable intersection control type where two roads cross?
- What is the most appropriate configuration for an intersection?
- How much lane capacity is adequate?
- What geometry is suitable (e.g. grade separate or at grade)?
- How will a particular intersection operate in terms of Level of Service (LOS) and delay in a future scenario given a certain volume of traffic?

When intersections are considered in isolation (where interaction between intersections are not important for the purpose of the exercise), deterministic models such as SIDRA, Oscady, Picady and Arcady are used. These models are reasonable inexpensive to operate and data input requirements are limited. Existing and expected traffic volumes at an intersection can be modelled assuming different intersection types and lanes or reconfigurations can be added until a geometrical configuration is obtained that adheres to acceptable design and operational standards. The model output in the form of LOS, vehicle delay and queue lengths will inform how to conceptually design an intersection to a geometrical optimal layout. A variety of Traffic Engineering Manuals and guidelines exist defining the appropriate and or acceptable requirements and specification to render the intersections' operating level acceptable.

Typical requirements and specification include limited vehicle delay, limited vehicle queuing and a Volume/Capacity ratio that is less than 0.95 (see Fig. 10).

Finding the appropriate intersection control type is an iterative process. The process that would be typically followed is to test the most simplistic intersection type first given a specific volume of traffic. A "build up" process is then followed. For example, the following intersection types could be assessed sequentially for optimal suitability until an intersection type is identified that satisfies guidelines, specifications and requirements:

- Give-way
- Three or four-way stop
- Roundabout/traffic circle
- Signalisation
- Grade separation

As each intersection type is considered, intersection specification is changed until an optimal layout/geometry is determined. For example, lanes can be added to accommodate additional through or turning movements. Traffic circles or roundabouts however presents issues of its

Intersection Type	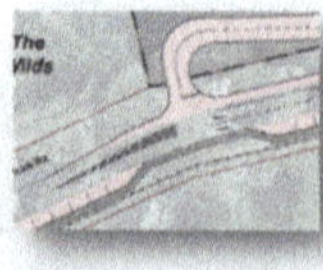 **Priority Intersection**	 **Traffic Signal Control**	 **Roundabout**
Specifications	Number of lanes Length of lanes Turning movements	Number of lanes Length of lanes Turning movements Signal phasing and length	Number of lanes Length of lanes Diameter size Gap acceptance

Figure 10: Requirements and specification for different intersection control types.

own including large land or space requirements and pedestrians having difficulty crossing from one side of the road to the other. Another factor to be considered is the cost of implementing the intersections. The more complex the intersection the higher the cost. Therefore, the decision about which intersection is the most appropriate depends on a variety of factors that cannot be fully addressed by transportation modelling alone. The most suitable intersection type is often a result of balancing different needs, consideration and stakeholder requirements.

Case study 4

The previous case study focused on defining the most appropriate intersection control type and geometrical layout of an intersection. However, intersections form part of a road network comprising a variety of roads intersecting with one another and their impact/influence on each other. Traffic flow at one intersection influences traffic flow at the next intersection (particularly in urban areas) and therefore it may not be realistic to design or assess intersections in isolation (particularly when the objective is to define the types of intersection control for a large area). Signal controlled intersections provides the opportunity to synchronise the operation of a series of intersections with one another allowing a platoon of vehicles arriving at a certain speed to be able to reach each intersection during the green phase and therefore not having to stop at any of the intersections. This is made possible by including an offset time equal to the anticipated time that it takes to travel from one intersection to another in the entire traffic phase.

The modelling software used for this operation would therefore need to be able to link one intersection to another and deterministic models that views intersections in isolation would not be able to provide this function. Fig. 11 illustrates a progression analysis output diagram for a series of three intersections from Transyt software. Transyt is particularly useful in signal optimisation and is reasonably inexpensive and data input requirements are limited. Typical data input requirements include intersections count data, queue length data and signal phasing plans.

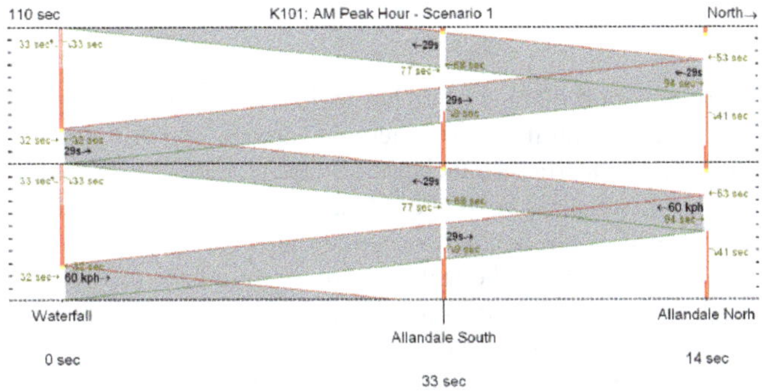

Figure 11: Transyt output.

Case study 5

In nodes or business districts where significant development growth occurs on multiple sites, a network based approach to transport planning is considered. The transportation model used should be able to assign traffic to the road network based on an estimated travel cost over a sufficiently wide enough area. The development of such a model includes an iterative process requiring multiple computations, which is too time consuming and repetitive to be calculated on paper or by using a spreadsheet.

For large-scale development or when having to present a region or urban node by a transportation model consisting of a vast network of road links and intersections affecting being interconnected and affecting one another a microscopic model is suitable, e.g. SATURN. These model types are expensive to develop, are traffic data hungry and take a lot of time to calibrate and validate.

The model would be developed as described in section 8. For example, the area would be divided into travel analysis zones and for each the number of trips ending in it or generated by it would be identified. Road links and intersection capacity will be coded and act as the potential pathways for trips to find their destination. The model will assign trips to the route that shows the least resistance or in other words takes the least time for a trip to be made between two points or nodes. The images below shows assigned traffic volumes for modelled areas in Johannesburg using SATURN software (see Fig. 12).

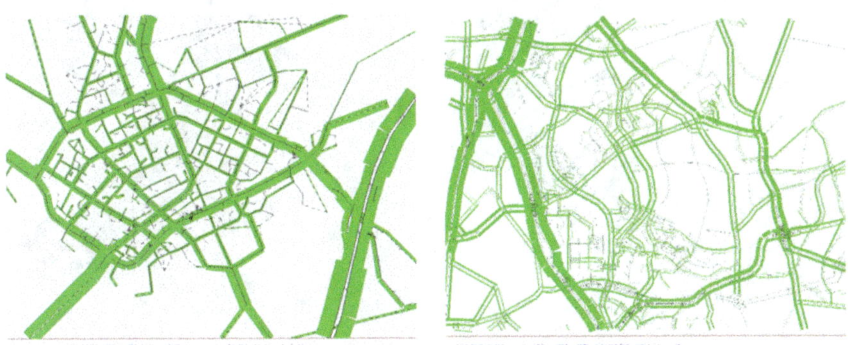

Figure 12: SATURN output.

Case study 6

The traffic engineer or transport planner often has to decide what transportation infrastructure would be most beneficial within a road network given specific traffic volumes and characterises. Large scale infrastructure such as new interchanges, major roads and or bridges has far reaching traffic impacts and costs. For instance, drivers living up to say 15 to 20 km from a major road network might decide to change their travel behaviour due to new road infrastructure provided elsewhere or the lack of it to improve their travel mobility.

Fig. 13 illustrates large scale infrastructure proposed at an existing grade separated interchange. Again, microscopic modelling was carried out using SATURN software to determine which of a number of geometrical options would produce the optimal level of benefit to the surrounding area and whilst maintaining the mobility function of the road network. In Fig. 13 below it was considered that the provision of the southern interchange provides a new route option, while the northern interchange only increases existing capacity on an established route. The new route option dramatically changes travel patterns and relieves road infrastructure currently operating under tremendous pressure. This infrastructure was considered to be the most beneficial and was recommended for implementation.

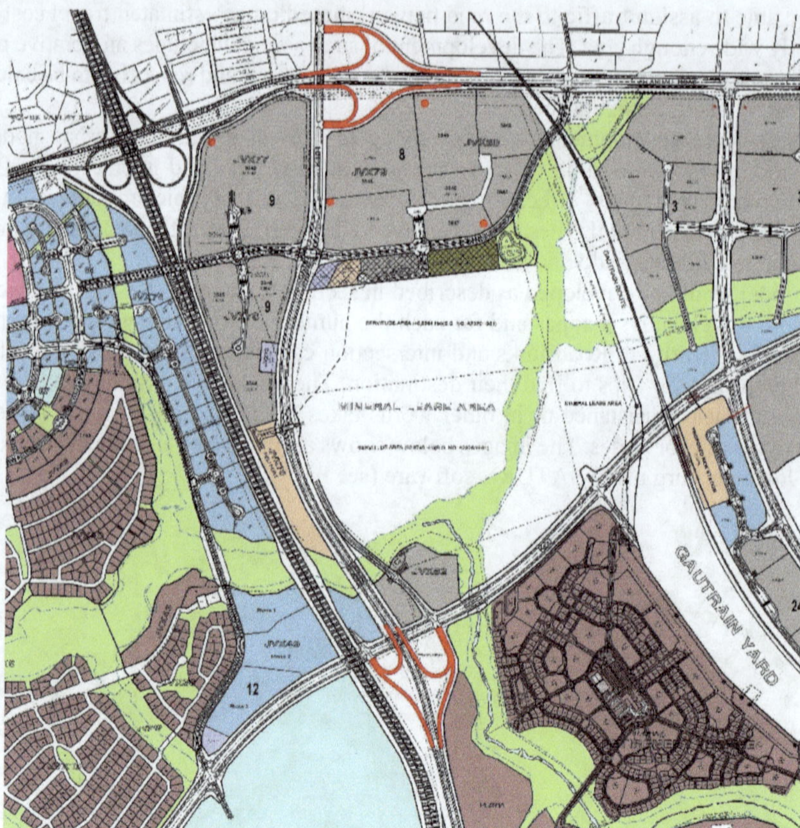

Figure 13: Large-scale infrastructure assessment.

Case study 7

At a strategic planning level decisions have to be made on regional, provincial and national levels to determine the impact of long distance routes and other transport infrastructure and the interaction between them. The type of route infrastructure and possible public transport systems involved has huge capital costs and the benefits of its provision need to be understood well by decision-makers. To support the decision-making process travel demand models are often used to demonstrate the future expected utilisation of proposed infrastructure. Such benefits are converted into social benefit i.e. travel time savings, new employment opportunities and development opportunities to help decide among different options.

Fig. 14 represents a snapshot of an EMME model, a microscopic travel demand model. Whilst SATURN reviews the road network in more detail by considering for example number of lanes on road links and exact intersection configurations, EMME considers each road only as a route option with a generalised capacity. It therefore determines long distance travel desire lines within a region and allocates traffic accordingly but also considering the routes of least resistance and is therefore employed given its strategic geographical nature advantage.

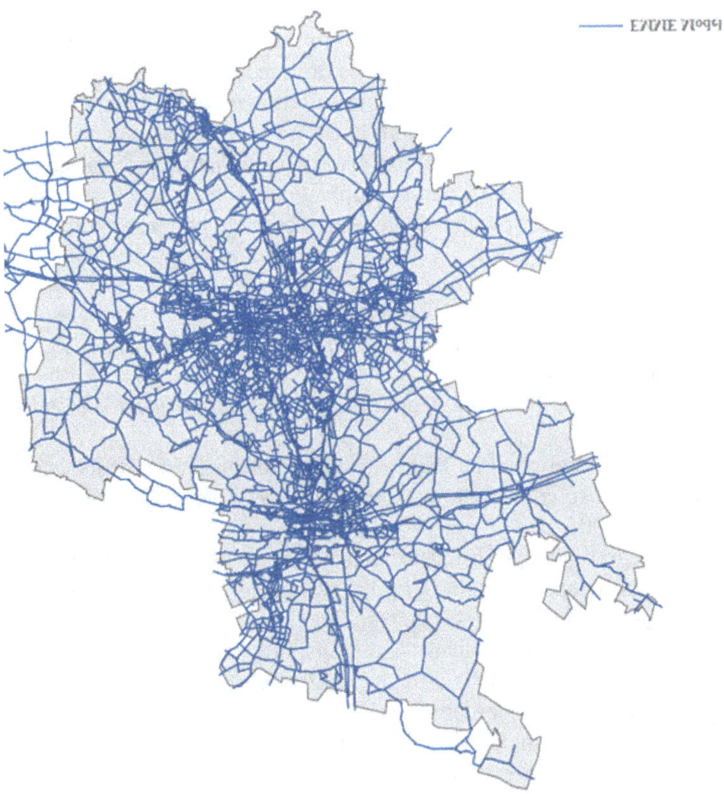

Figure 14: Regional transport model.

11 CONCLUDING COMMENTS

Transportation modelling and simulation as interface in the planning process plays an important role in enabling spatial development. The models developed by transport planning professionals are simplified representations of reality.

Transportation modelling is a young and developing science that has generally developed as part of the evolution of advancing computer technology and was originally developed to solve highway capacity problems. Many mistakes have been made in the past where inappropriate solutions were implemented – with hind sight. A classic mistake includes where new infrastructure provision is considered purely on the basis of the "predict and provide" approach resulting in infrastructure being implemented completely surplus to requirements and unnecessary. The consequence is an unpractical and unaffordable cost to government.

However, although mistakes have been made in the past, models have also proven their worth in solving highly complex problems and the current diverse range of modelling software available means that models are becoming more bespoke in how they support traffic engineers and transportation modellers addressing specific problems.

Transportation models should be seen for what they are: a support tool that helps in decision making. They will always require humans to contextualise and interpret outputs, considering the results in context of specific environments with unique characteristics, features, balancing needs, different stakeholder opinions and humans occupying and living in space.

Over time the relationship between transportation and land use has been better understood. Transport design more often than not determines the success of urban environments and can either be an enabling factor or a hindrance. Transportation modelling should therefore always be considered as a tool that supports decisions and not the panacea to every traffic – transport – land use problem we face or wish to solve. It is expected that transportation modelling will diversify and develop significantly more sophisticated techniques to simulate human behaviour and hence transport flow in the decades to come to better support decision making in planning.

REFERENCES

[1] Goulias, K.G., *Travel behavior and demand analysis and prediction*, University of California, 65 pp., 2007.

[2] Tri-County Regional Planning Commission, *Travel forecast model development*, 14 pp., 2009.

[3] American Association of State Highway and Transportation Officials, *Handbook on integrating land use considerations into transportation projects to address induced growth*, 52 pp., 2005.

[4] Toth, G., Integrating land use and transportation planning through placemaking. Projects for public space, Online. http://www.pps.org/blog/integrating-land-use-and-transportation-planning-through-placemaking/

[5] Wegener, M., Overview of land use transport models. *Handbook of Transport Geography and Spatial Systems, Vol. 5, Handbooks in Transport*, eds D.A. Hensher & K.J. Button, Elsevier: Amsterdam, pp. 127–146, 2004.

[6] United States Department of Transportation Federal Highway Authority of Transportation, Planning for transportation in rural areas, Online. http://www.fhwa.dot.gov/planning/publications/rural_areas_planning/page06.cfm, 2012.

[7] United States Department of Transportation Federal Highway Administration, The transportation planning process: key issues, 77 pp., 2007, Online. https://www.planning.dot.gov/documents/briefingbook/bbook.htm

[8] Victoria Transport Policy Institute, TDM Encyclopaedia. Transport model improvements improving methods for evaluating the effects and value of transportation system changes, 2014, Online. http://www.vtpi.org/tdm/tdm125.htm

[9] Litman, T., Understanding transport demands and elasticities: How prices and other factors affect travel behaviour, Victoria Transport Policy Institute, 76 pp., 2013.

[10] Chandrasekaran, B., Survey of network traffic models, Online. http://www.cse.wustl.edu/~jain/cse567-06/traffic_models3.htm

[11] Florian, M., Models and software for urban and regional transportation planning, Centre for Research on Transportation, 2008.

[12] Brockwell, A.E., Chan, N.H. & Lee, P.K., A class of models for aggregated traffic volume time series, 17 pp., 2008.

[13] Jha, K., Sinha, N., Arkatkar, S.S. & Sarkar, A.K., Modelling growth trend and forecasting techniques for vehicular population in India. *International Journal for Traffic and Transport Engineering*, **3**(2), pp. 139–158, 2013.

[14] Meyer, D.M. & Miller, E.J., *Urban Transportation Planning: A Decision-Oriented Approach*, McGraw-Hill: USA, 2001.

[15] He, N. & Zhao, S., Discussion of basic elasticity models applied to induced traffic in China. *Proceedings of the Eastern Asia Society for Transportation Studies*, **9**, 2013.

[16] Boxill, S.A. & Yu, L., An evaluation of traffic simulation models for supporting ITS development, Center for Transportation Training and Research, Texas Southern University, Houston, 116 pp., 2000.

[17] Department of Transport, *Revised South African Trip Generation Rates Manual*, 2007.

[18] Committee of Transport Officials, *Technical Methods for Highways (TMH) 17, Vol. 1, Trip Data Manual*, 84 pp., 2012.

[19] Institute of Transport Engineers (ITE), *Trip Generation Manual*, 9th ed., 1216 pp., 2012.

[20] Chiu, Y., Bottom, J., Mahut, M., Paz, A., Balakrishna, R., Waller, T. & Hicks, J., *Dynamic Traffic Assignment*, Transportation Network Modelling Committee, Transportation Research Board, 62 pp., 2011.

[21] United States Department of Transportation Federal Highway Administration, Calibration and Validation, 36 pp., 2009.

[22] Wegmann, F. & Everett, J., *Minimum travel demand model calibration and validation guidelines for the State of Tennessee*, University of Tennessee, Center for Transportation Research, 25 pp., 2009.

NETWORK DESIGN AND CHANNEL STRATEGIES

GERARD DE VILLIERS
Arup (Pty) Ltd, Johannesburg, South Africa

ABSTRACT

Supply chains compete and not products, commodities or companies. Transport is of particular importance in logistics and supply chain management as transport costs are the largest cost component and probably the most important logistics function in the supply chain. The design of logistics networks and the development of appropriate channel strategies should consequently be an integral part of transportation planning and modelling.

This chapter starts with an introduction that explains the importance of network design and channel strategies in transportation modelling. It continues with a description of logistics and supply chain management and follows with the importance of the planning hierarchy. The next part covers the distribution channel and the different alternative options with practical examples. It also includes the implication of using intermediaries in the distribution channel as well as the role of third-party and fourth-party logistics service providers.

Location analysis covers some of the relevant theories on macro level as well as specific questions to be addressed on micro level. Logistics network design explains the trade-offs in determining the number of facilities and it also suggests typical outbound distribution strategies that are used.

The next part covers centre-of-gravity analysis. This type of heuristic modelling is done with spreadsheets without dedicated optimisation software. Mathematical formulae for the optimisation of the locations are explained and a practical example was included showing the calculations in tabular and graphical format. This is followed with a discussion on total logistics costs.

This chapter ends with reference to different channel strategies once the location and total logistics costs have been determined.

Keywords: centre-of-gravity analysis, channel strategy, distribution channel, fourth-party logistics, location analysis, logistics, network design, planning hierarchy, supply chain management, third-party logistics, and total logistics costs.

1 INTRODUCTION

The design of a logistics network and the configuration of the appropriate supply chain strategy to serve the network, are important prerequisites for the successful development of efficient supply chains. Christopher [1] suggests that leading-edge companies realised that the real competition is not company against company but rather supply chain against supply chain. This implies that suitable network design and channel strategies should receive due attention in the transportation modelling stage of the development of master plans and other development frameworks.

Cost of logistics is another key performance indictor of the supply chain that has to be carefully monitored and managed to be able to allow the supply chains to compete effectively. The CSIR [2] found in their latest annual State of Logistics Survey for South Africa that total logistics costs as a percentage of GDP remained constant at 12.5% since 2011. However, due primarily to fuel inflation, they expect a small increase to 12.8% in 2014. This remains considerably higher than the figure of 8.2% in 2013 for the USA [3]. A breakdown of the components of the latest logistics costs in South Africa is estimated for 2014:

- Transport: 58.8%
- Inventory carrying costs: 13.5%

- Management and administration: 12.4%
- Warehousing: 12.3%

The transport costs of 58.8% are exceptionally higher that the global average of 39%, but can partially be explained by the spatially challenged economy, largely driven by fuel and wages. The high transport cost confirms the need for careful planning for transportation and hence explains the importance of efficient network design and appropriate channel strategies.

2 LOGISTICS AND SUPPLY CHAIN MANAGEMENT

Various definitions of logistics and supply chain management are applied in practice and it might be useful to use a diagram to explain the concepts as used for purpose of this discussion. This is presented in Fig. 1 and indicates a flow of materials, products, commodities and cargo (collectively called "goods") from the supplier (inbound) side to the customer (outbound) side.

Raw materials are sourced and enter the supply chain on the inbound side and finished products or services exit the supply chain on the outbound side, with goods moving through the supply chain from origin to destination and information moving both forward and backwards through the supply chain. Reverse logistics moves upstream and includes goods returned, value carriers returned (such as pallets and empty gas bottles), disposal of assets, recycling materials, packaging materials and waste.

Transport plays an important part throughout the supply chain and from a simplified perspective, starts with bulk transport modes such as rail and sea on the inbound side, break-bulk transport in large trucks in between the plant and distribution centre (DC) and different size road transport vehicles carrying goods in unitised format (pallets or cartons) on the outbound side to final destination.

The inbound side used to be called materials management (now supply management) and the outbound side physical distribution (now distribution management), while the two sides together is called logistics management. The logistics elements or functions active *en route* from origin to destination include:

- Transport;
- Warehousing;
- Inventory management;
- Procurement;
- Customer service;
- Materials handling;
- Reverse logistics;
- Order management;
- Packaging;
- Freight forwarding; and
- Information systems.

Supply chain management on the other hand, implies the integration of all the mentioned elements or functions from origin to destination to provide for the management of a coherent sequence of activities from the suppliers of your suppliers to the customers of your customers and eventually to the consumers in the case of consumable products. This implies that a number of individual supply chains as indicated in Fig. 1 are all being integrated and interlinked into a complex network of demand and supply that requires very special management skills.

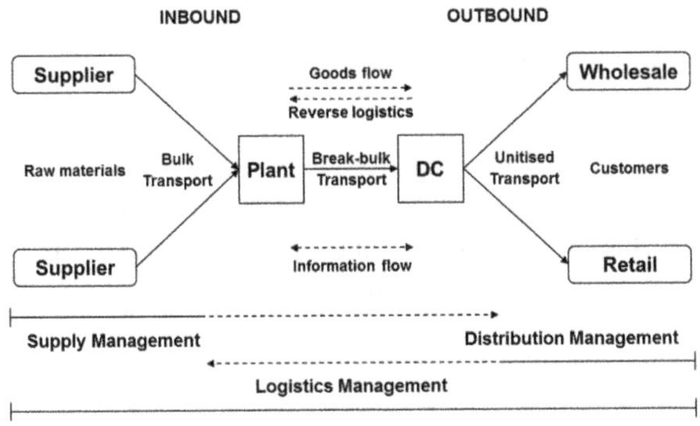

Figure 1: Logistics management (adapted from De Villiers et al. [4]).

Christopher [5] discusses the future requirement for structural flexibility in this complex environment and suggests that as supply chains become more 'virtual' than 'vertical', there will be a growing need for network orchestration. The ability to structure appropriate networks and to synchronise the activities across the nodes and links of those networks will be paramount in the era of network competition.

3 PLANNING HIERARCHY

Stock and Lambert [6] use the planning hierarchy (originally developed by Copacino of Andersen Consulting in 2000) very effectively to develop a strategic logistics plan. Fig. 2 provides an adapted version of this planning hierarchy to include *inter alia* procurement as well as freight forwarding on the functional level, plus some minor editing to expand description of some of the key areas.

Planning should start at the strategic level, and follows in a specific order through the structural, functional and operational levels as indicated by the arrow, to ensure structure follows strategy. Conceptual issues are addressed at the strategic and structural levels, with an overlap into the functional level, while detail issues receive attention on the functional and operational levels. Assumptions are allowed on conceptual level where less accuracy is needed but calculations made on the detail level need to be more accurate.

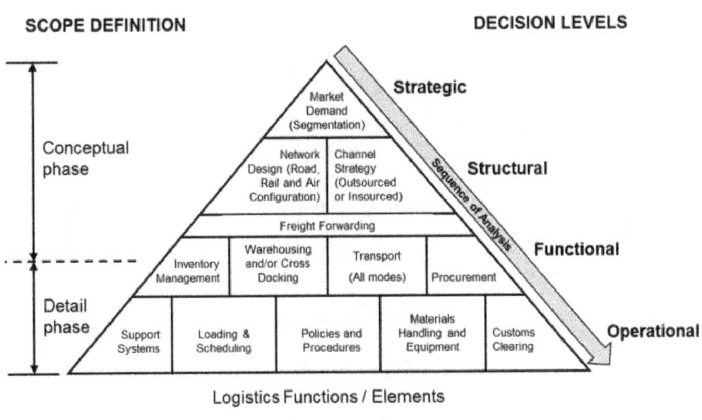

Figure 2: Planning hierarchy (adapted from Stock and Lambert [6]).

The pinnacle of the triangle or hierarchy starts with the specific requirements of the customer, based on an analysis of the market demand in specific segments as well as the anticipated growth in demand. Once the demand is determined and future growth estimated, the planning process should proceed to network design and channel strategy on structural level.

Network design addresses questions such as the number and location of facilities (including plants, distribution centres, warehouses, depots and terminals) and centralisation *versus* decentralisation while channel strategy refers to roles and responsibilities of stakeholders and strategies such as outsourcing, insourcing and combinations thereof. Once the network design and channel strategy are done, planning moves to the functional level where more detail of the respective logistics elements or functions such as inventory management, warehousing, transport and procurement are addressed. This includes inventory replenishment strategies, warehouse automation, transport mode and type selection and centralised *versus* decentralised purchasing.

Once the functionality of the elements are known, specific design of support systems, loading and scheduling technology, policies and procedures, materials handling and customs procedures can be addressed on operational level.

So, as example, one can only address racking and shelving (materials handling and equipment) once the warehouse design has been done, after the decisions were made regarding centralised or decentralised warehousing on structural level (network design), which can only be done once demand has been confirmed on strategic level.

The importance of the planning hierarchy should not be underestimated as it provides a very useful framework for strategic planning in general, and transportation planning in particular.

4 DISTRIBUTION CHANNEL

Distribution management was defined in Fig. 1 as the outbound side mainly from the DC to the customer, although sometimes the link from the plant to the DC is included. Distribution includes both transport and warehousing and distribution channels refer to the path that the goods travel outbound to final destination. Some of the various options for this path are provided in Fig. 3.

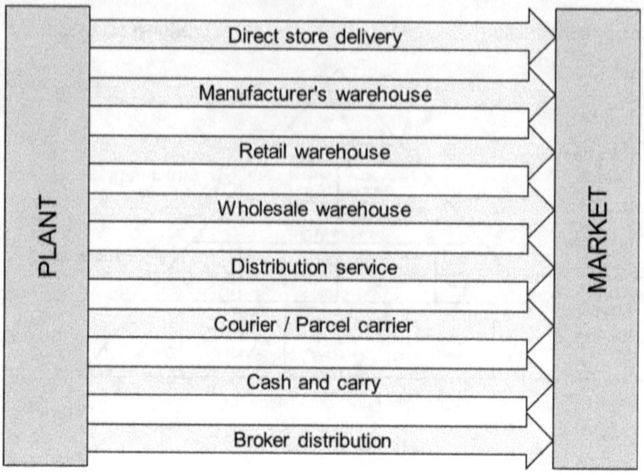

Figure 3: Alternative distribution channels (adapted from Rushton et al. [7]).

Each of the respective distribution channels demands specific design criteria for activities such as:

- Inventory management (such as cycle stock and safety stock levels);
- Replenishment strategy (such as period review and min-max levels);
- Warehousing (such as centralisation versus decentralisation);
- Load and drop sizes;
- Transport mode and type;
- Fleet type and size;
- Outsourcing versus insourcing for various functions; and
- Service levels.

The choice of distribution channel is further impacted by the characteristics of goods handled in the channel as well as the format of handling. Dry bulks is different from liquid bulk; break-bulk in containers or on pallets require completely different handling and the cold chain for chilled or frozen goods has very specific and expensive requirements to ensure integrity of temperature along the distribution channel.

Rushton et al. [7] are of the opinion that channel objectives will necessarily differ from company to company but is it important to intentionally agree on the objectives as the design of the distribution channel is directly linked to the objectives. In addition to the objective of maintaining the cold chain intact as mentioned, they suggest the following possible objectives:

- To make the product readily available to the market at which it is aimed;
- To enhance the prospect of sales being made;
- To achieve cooperation with regard to any relevant distribution factors;
- To achieve a given level of service;
- To minimise logistics and total costs; and
- To receive fast and accurate feedback of information.

Fig. 4 provides an indication of alternative options for conventional fresh fruit distribution channels when exporting through a deep-sea port to foreign markets. The options include various combinations of ambient and refrigerated transport, accumulating buffer stock at the port in anticipation of the arrival of the vessel, different cooling options from cooling at a centralised facility to cooling at decentralised depots, and finally loading options at the port *en route* to the market.

Similar options exist for the export of fresh fruit in containers and that is shown in Fig. 5. The location of stuffing the containers (packing the pallets into the containers) can be at a centralised facility, at decentralise depots or at the port. Latest trends favour containerisation to most developed countries where ports are designed for container handling and no break-bulk or palletised goods will be handled, but this is not the case in developing countries where various ports in Africa, as example, prefer labour-intensive handling. This is due to a lack of suitable container handling equipment or where affordable manual labour is available.

In both examples of conventional and containerised handling, the alternative distribution channels show the option of a centralised facility at the centre of gravity (CoG). This will be discussed in detail in Section 7 but the role of intermediaries such as this centralised facility requires some attention before we move on.

Lambert and Stock [8] provide a practical example of how intermediaries can reduce the cost of contact between supplier and customer. In the case of direct distribution from four suppliers to ten customers, the contacts are 4 x 10 or 40 contacts in total. This is shown in Fig. 6.

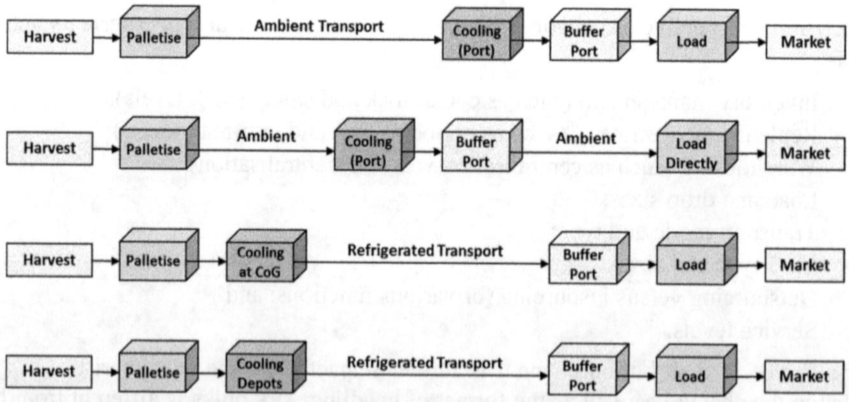

Figure 4: Conventional fresh fruit distribution channels.

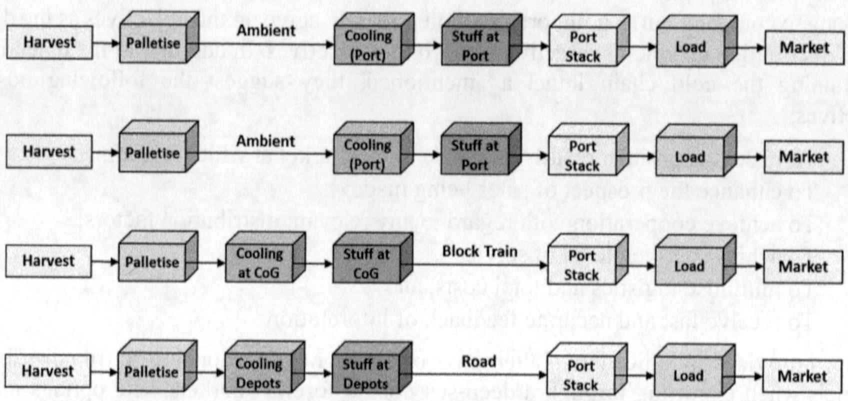

Figure 5: Containerised fresh fruit distribution channels.

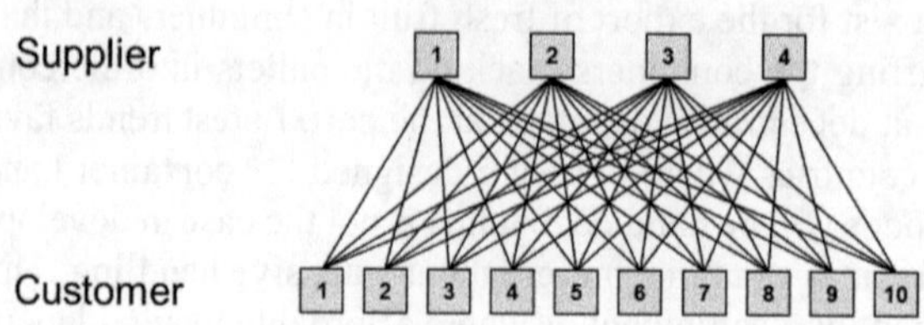

Figure 6: Direct distribution (40 contacts) [8].

However, in the case of distribution through an intermediary from four suppliers to ten customers, the contacts are 4 + 10 or 14 contacts in total, as shown in Fig. 7. This is significant less than the 40 contacts in direct distribution and provides much less interruption along the channel.

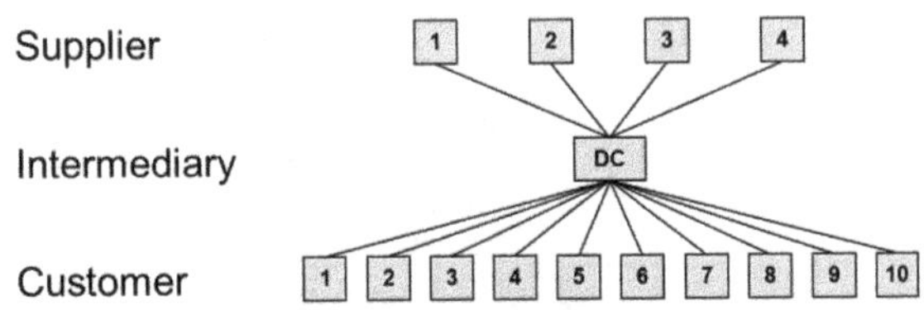

Figure 7: Distribution through intermediary (14 contacts) [8].

The role of intermediaries depends on the specific logistics function and could include managing a warehouse or distribution centre as shown in the figures or it could be taking responsibility for transport. This is called outsourcing and implies a third party moves in between the supplier (first party) and the customer (second party), as shown in Fig. 8. The first party is usually the seller, the second party is the buyer and the third party is the service provider in the middle. The third party is called a "third-party logistics service provider" (3PL) and operates on behalf of either the supplier or the customer.

Industry experience has confirmed that unfortunately not all customers are satisfied with the 3PL services due to some sub-optimal rendering of services which could have been integrated and offered at lower cost to the supply chain. An example could be the export of chemicals by one company and the import of chemicals by another. Often the two companies are served by two service providers while the liquid bulk carriers should rather be optimised with the same vehicles used for the export and imports, preventing empty trucks and full trucks to pass each other on the same route and in the same channel.

This has caused some companies to reconsider outsourcing and rather bring the fleet back in-house (insourcing) to be able to manage the function more optimal in house. Gattorna [10] introduces the concept of "fourth-party logistics service provider" (4PL) as a possible solution that incorporates the advantages of both outsourcing and insourcing. A new entity is established to manage multiple 3PL service providers as indicated in Fig. 9, to achieve much better integration and improved service delivery at much lower cost.

According to Gattorna [10], the 4PL functions as "an integrator that assembles the resources, capabilities and technology of its own organisation and other organisations to design, build and run comprehensive supply chain solutions and which have the cultural sensitivity, political and communication skills, and the commercial acumen, not only to find value, but to create motivating and sustainable deals that offer incentives to all parties involved".

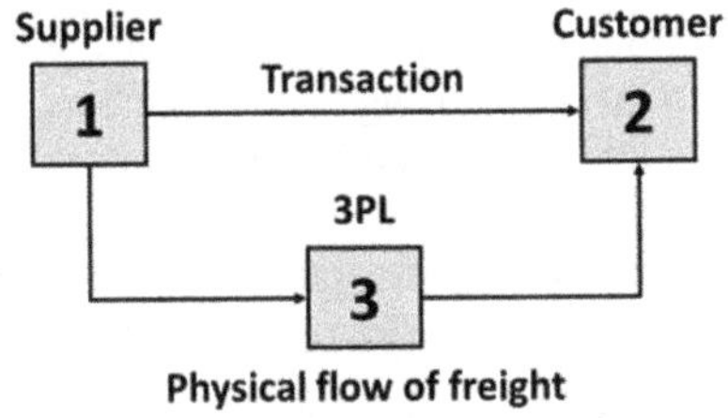

Figure 8: Third-party logistics (3PL) [9].

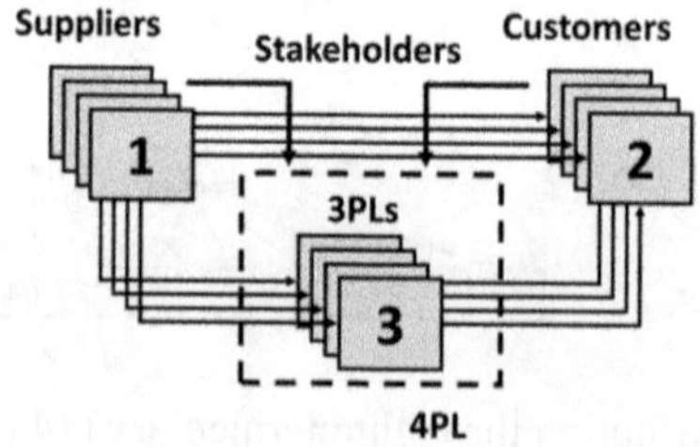

Figure 9: Fourth-party logistics (4PL) [9].

In essence, this means a solution that incorporates the advantages of both outsourcing and insourcing to provide maximum overall benefit. It differs from traditional 3PL arrangements in four main respects:

- The 4PL organisation is often a separate entity established as a Joint Venture (JV) or long-term contract between a primary client and one or more partners;
- It acts as a single interface between the client and multiple logistics service providers;
- All aspects of the client's supply chain are managed by the 4PL organisation; and
- It is also possible for a major third-party logistics provider to form a 4PL organisation within its existing structure.

The 4PL concept addresses the shortfalls of traditional 3PL arrangements and offers the opportunity to achieve substantial incremental benefits. In addition, the concept can be extended to an existing 3PL, converting it into a 4PL.

5 LOCATION ANALYSIS

Stock and Lambert [11] provide a useful overview on how to determine the best place for a warehouse and they suggest the following three general locations from a macro perspective:

- market positioned warehouse;
- production positioned warehouse; and
- intermediately positioned warehouse.

The first option of market positioned, locating the facility closest to the final customer, should be considered for service sensitive distribution channels and where close proximity to the customer is crucial. This is also true for the location of a production facility (or plant) for "weight-gaining" goods such as bottling of beverages. The second option of production positioned, locating the facility close to the sources of supply or production function, should be considered for perishable products where considerations other than customer service are important. This also applies to "weight-losing" goods such as iron ore converted into steel. The third location is obviously a hybrid of the first two and will be determined by trade-offs between service and cost.

The authors continue with a discussion on macro considerations and they include reference to the most relevant location theories of the following economic geographers:

1. Johann Von Thunen, a German agriculturist, suggested that transport cost minimisation was the most important locational determinant. Agricultural production would take place where the farmer would maximise profits and as locations further

from the city (market) would incur greater transportation cost, low-value products should be produced near the city.

2. Alfred Weber, a German economist, developed a theory for the location of industrial production facilities. He assumed equally accessible transportation and constant transportation costs with respect to weight and distance, and defined the optimum location as the point that represents the least-cost location. More specifically, the least-cost site is the location that minimises total transportation costs; for both raw materials and finished goods to the market.
3. Edgar Hoover, an American theorist, investigated the optimum location of industrial facilities based on cost factors, including demand factors of the market area after a firm determines a location. He considered transportation costs, agglomerative factors and industrial costs. His analysis was based on the fact that transport costs are not linear with respect to distance.
4. Melvin Greenhut expanded on the work of the three mentioned geographers and included environmental and security factors as well as profitability elements in the location choice. He suggests the optimal facility location to be where profits are maximised.

Ackerman [12] suggests that once the macro analysis provided a suitable region, attention should be given to the micro level factors and he recommends the checklist of questions indicated in Table 1.

The list of questions seems daunting but it is important that the location analysis is done properly from the start to prevent surprises and expensive requirements for compliance once the project started. It might be a good idea to do the location analysis (both macro and micro level) on a reasonably high level to allow further refinements once the logistics network design has been done and the centre of gravity located. This implies an iterative process that will yield the best results and highest possibility of choosing the most optimal site.

6 LOGISTICS NETWORK DESIGN

Logistics network design follows in the planning hierarchy directly behind market demand and segmentation and after looking at macro level at possible regions for the development of one centralised distribution centre or a network of facilities for decentralised distribution.

Simchi-Levi et al. [13] suggest that network design determines the physical configuration and infrastructure of the supply chain. This is a strategic decision that has long-lasting effect on the firm and involves typical decisions related to plant and warehouse location as well as distribution and sourcing. The objective is to design or configure the logistics network to minimise production costs, purchasing costs, inventory holding costs, facility costs and transportation costs. The authors are of the opinion that the trade-offs are clear and explains that increasing the number of warehouses typically yields:

- An improvement in service level due to a reduction in average travel time to the customers;
- An increase in inventory costs due to increased safety stocks required to protect each warehouse against uncertainties in customer demands;
- An increase in overhead and setup costs;
- A reduction in outbound transportation costs (from warehouses to customers); and
- An increase in inbound transportation costs (from the suppliers and/or manufacturers to the warehouses).

Table 1: Check list to locate your next warehouse (adapted from Ackerman [12]).

Category	Question
Government restrictions	1) Is the property properly zoned for warehousing or possible future uses? 2) Are there limits or restrictions on parking? 3) What is the maximum expansion capability? 4) Does the property have any easements or protective covenants? 5) Do access roads have any restrictive load limits?
Geographic restrictions	1) It the grade too steep for construction? 2) Are there drainage problems? 3) Are there streams or wetlands near the site? 4) Where is the 100-year floodplain? 5) Do have geodetic survey information on seismic fault lines? 6) Has the site be used for landfill and is there landfill debris on the site? 7) What is the load bearing capacity of the soil? 8) Are there restrictions on vegetation removal? 9) Are there any environmental fatal flaws on the site?
Transportation	1) What is the access to the nearest highway interchange? 2) What is the distance from road freight terminals? 3) Have we interviewed road freight managers about this site? 4) What is the cartage cost to the closest inland container terminal? 5) What is the cartage cost to the closest maritime container terminal? 6) What is the cartage cost to the closest air freight terminals? 7) Is the site convenient for pickups by our most important customers?
Utilities	1) What is the source of water supply? 2) Is the volume and pressure of water sufficient to support an early suppression fast response (ESFR) sprinkler system? 3) What is the electric power supply? 4) Is it possible to install alternative energy technology to mitigate power interruptions? 5) Have we received recent utility bills? 6) What incentives are available from the electric utility supplier? 7) Is natural gas available? 8) Is sub-metering or prepaid electric and gas permitted? 9) Are there any limitations on utility capacity or demand?
Security	1) Does the sprinkler system meet preferred risk standards? 2) Is there a secondary water source for the sprinkler system? 3) Have you obtained recent insurance inspection reports? 4) Have you interviewed law enforcement officials? 5) Have you interviewed managers of neighbouring facilities? 6) Are armed response services available?
Labour	1) Are nearby businesses unionised? 2) What is the community's attitude towards trade unions? 3) Is there a labour shortage or high unemployment? 4) What is the quality rating of nearby schools?
Community	1) Do local civic leaders welcome industrial expansion? 2) Is there any opposition to warehouse operations? 3) Is there a history of difficulty in rezoning?
Taxation	1) Have you obtained the most recent property appraisal? 2) Have you analysed real estate tax history over the past 10 years? 3) Have there been any recent tax assessments? 4) Are there any tax abatement programs available? 5) Is the location in a duty-free or enterprise zone (Industrial Development Zone or Special Economic Zone)? 6) Have you obtained the plans for any public improvements? 7) How do inventory tax levels compare with other sites? 8) Have you obtained all of the appropriate tax rate information?
New construction	1) Have soil tests be done? (Geotechnical fatal flaw analysis) 2) Are their landscaping requirements? 3) Have you done an Environmental Impact Assessment? 4) What are the estimated costs and time requirements for permits and development approval?

It is clear that the answer lies in total logistics costs, and not individual transport, warehousing or inventory costs. This will be discussed in more detail in Section 9 but suffice it to say that Blanchard [14] calls this "striking the proper balance". He quotes Harmelink as saying that "the objective of strategic network planning is to come up with the most economical way to ship and receive products while maintaining or increasing customer satisfaction requirements; simply put, a plan to maximise profits and optimise service".

The alternative different distribution channels as shown in Fig. 3 assume the existence of distribution centres, warehouses and or depots, collectively referred to as storage facilities (or nodes in the links between origin and destination). These facilities are important components of any logistics system and most channel strategies rely on some intermediate break-bulk and storage facility between producer and retailer or consumer.

Blanchard [14] lists a number of questions to be answered in developing a suitable network, and Table 2 links the questions to the specific level of the planning hierarchy as discussed in Section 3. It is clear that the questions are relevant at and should be addressed on a number of the levels in the planning hierarchy.

In conclusion, Simchi-Levi et al. [15] suggest that typically three distinct outbound distribution strategies (or network design options) are used:

- Direct shipment with items shipped directly from the supplier to the retail stores without going through distribution centres;
- Warehousing as the classical strategy in which warehouses keep stock and provide customers with items as required; and
- Cross docking where items are distributed continuously from suppliers through warehouses to customers. However, the warehouses rarely keep the items for more than 10 to 15 hours.

The choice between those options (and others) can be very complicated, with detail analyses required to establish the cost trade-offs, as well as quantitative techniques to determine storage facility location. There might not necessarily be one best solution but an integrated system of nodes and links (facilities and channels) in the logistics network.

One technique for solving the complicated choice is centre-of-gravity analysis, which is probably the most powerful tool for designing a new network from the basis of a "clean slate''. It enables an objective assessment of the actual centres of demand and supply, as well as of the combined position of both supply and demand. This will be discussed in the next section.

Table 2: Network design questions.

Question	**Planning hierarchy level**
1) How many distribution centres (DCs) do you need?	Structural level
2) Where should the DCs be located?	Structural level
3) How much inventory should be stocked at each DC?	Functional level
4) Which customers should be serviced by each DC?	Structural and functional level
5) How should customers order from the DC?	Operational level
6) How should the DCs order from the suppliers?	Operational level
7) How often should shipments be made to each customer?	Functional and operational level
8) What should the service levels be?	Strategic level
9) Which transportation methods should be used?	Functional level

7 CENTRE-OF-GRAVITY ANALYSIS TECHNIQUE

The design of a suitable logistics network and the development of appropriate channel strategies in multiple markets and multiple suppliers is a daunting task and some modelling is needed to assist with the complex calculations. Coyle et al. [16] calls centre-of-gravity analysis the "grid technique" and suggest that to be a useful, simplistic heuristic modelling approach for determining the least-cost facility location. This modelling can easily be done by spreadsheets and does not require dedicated optimisation software.

Centre-of-gravity analysis is used to determine the least-cost location of a fixed facility (plant, distribution centre, warehouse, depot or terminal) for moving raw materials and finished goods in the network. The technique assumes sources of raw materials and markets are fixed and that supply and demand are known. The coordinates of each supply and demand point are needed to present the points in geographical format on a map and to calculate the transport costs of all movements.

The technique is shown graphically in Fig. 10. If all the points are plotted at their respective coordinates on a map and respective supply or demand volumes (represented by weights) are connected through small holes to a central ring on the surface, the ring will come to rest in the centre of gravity.

The analysis is based on the differential transport costs of raw materials from the production or manufacturing facilities or suppliers to one or more fixed facility such as distribution centres (primary transport), and finished goods from these facilities directly to destinations in the market, such as retail outlets (secondary transport). The result of the analysis provides a theoretical indication of the least cost location of the centralised facility.

Bowersox and Closs [17] provide four analytical techniques for calculating the centre or gravity:

- Ton-centre solution (weight-centre solution);
- Mile-centre solution (distance-centre solution);
- Ton-mile-centre solution (weight-distance-centre solution); and
- Time-ton-mile-centre solution (time-weight-distance-centre solution).

The ton-centre solution (weight-centre solution) is the basic centre-of-gravity calculation and the centre of movement represents the least cost location. The mathematical formula is presented in eqn (1).

$$x = \frac{\sum_{i=1}^{n} x_i F_i}{\sum_{i=1}^{n} F_i}, \qquad y = \frac{\sum_{i=1}^{n} y_i F_i}{\sum_{i=1}^{n} F_i} \tag{1}$$

where:

x, y = Coordinates for the centre of gravity
x_i, y_i = Coordinates for all origins and destinations
F_i = Supply or demand (ton)

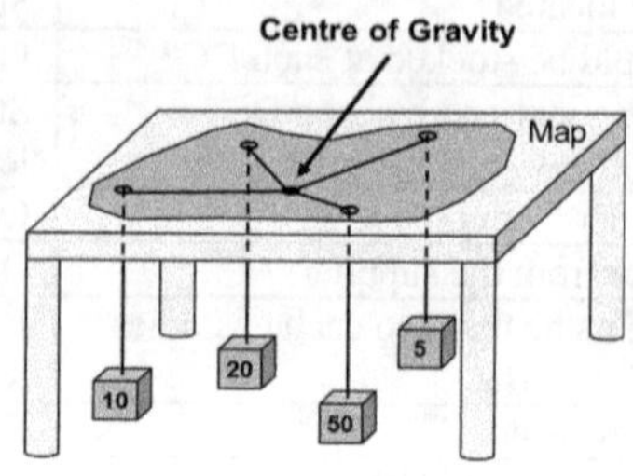

Figure 10: Centre-of-gravity technique.

The mile-centre solution (distance-centre solution) determines the geographical point that minimises the combined distance to all points.

The ton-mile and time-ton-mile-centre solutions reflect most accurately reality and should be used where possible. It can incorporate a number of variables that needs to be considered, such as transport rates (Rand/ton.km) which is different for primary and secondary transport.

Similarly, a differential parameter for congestion in urban areas that reflects the additional time required, can be used. This parameter could range from high value for high congestion to a low factor for low congestion. The mathematical formula of a hybrid between the two solutions (adapted from Bowersox and Closs [17]) is shown in eqn (2).

$$x_k = \frac{\sum_{i=1}^{n} x_i F_i R_i C_i / d_i}{\sum_{i=1}^{n} F_i R_i C_i / d_i}, \quad y_k = \frac{\sum_{i=1}^{n} y_i F_i R_i C_i / d_i}{\sum_{i=1}^{n} F_i R_i C_i / d_i}, \tag{2}$$

where:

x_k, y_k = Coordinates for iteration k (in km)
x_i, y_i = Coordinates for all origins and destinations (in km)
F_i = Supply or demand (ton)
R_i = Transport rate (R/ton.km)
C_i = Congestion factor (Relative value)
d_i = Distance between points for each iteration

This calculation requires an iterative process to determine the increasingly improved location, based on the least cost of transport. The distance between the locations, with each iteration reduces every time a better location is found and the objective is to found the spot where the costs are lowest. The measurement of distance between two points is done by the formula in eqn (3):

$$d_i = \sqrt{(x_i - x_k)^2 + (y_i - y_k)^2}, \tag{3}$$

where:

x_k, y_k = Coordinates for iteration k (in km)
x_i, y_i = Coordinates for all origins and destinations (in km)
d_i = Distance between points for each iteration

The calculations are relatively easily done by spreadsheets and are shown in tabular and graphical format in an example in the next section.

8 CENTRE-OF-GRAVITY ANALYSIS EXAMPLE

The application of the centre-of-gravity technique can be shown in the following hypothetical example; where the purpose is to determine the least-cost location of a new distribution centre.

The market consists of five customer destinations located in five different areas, with the weekly demand at each customer location as follows:

- Customer 1: 20-ton product
- Customer 2: 10-ton product
- Customer 3: 15-ton product
- Customer 4: 5-ton product
- Customer 5: 12-ton product

The supply to the market consists of two local suppliers and two foreign suppliers from where imports are received by air and by sea from the following four origins:

- Supplier 1: 11-ton product
- Supplier 2: 11-ton product
- Imports by sea: 20-ton product
- Imports by air: 20-ton product

Inbound (primary) transport from the local suppliers, the airport and the sea port is done by 20 ton trucks at a rate of R 0.60 per ton.km while outbound (secondary) transport or delivery to the customers is done by 8 ton trucks at a rate of R 1.20 per ton.km.

This information is presented in Table 3 with the respective calculations. The known values include volume (in ton) supply or demand at each origin and destination as well as the respective transport rates (in Rand/ton.km). The centres of gravity are consequently calculated for the demand centre of gravity, supply centre of gravity as well as the combined centre of gravity.

Optimisation of the location is done by an iterative process where new coordinates for the best location is calculated by sequential iterations until the difference becomes marginal. The optimisation iteration process of 15 iterations is shown in Table 4 and the final location also presented in Table 3.

The results of the calculations done in the two tables are provided graphically in Fig. 11.

The locations of the customers are indicated with circles, the location of the supply points with blocks and the location of the centres of gravity with diamonds. The supply centre of gravity is located closer to the airport and seaport where most products originate and the demand centre of gravity is somewhere in the middle of the market.

The combined centre of gravity is located on a straight line between the supply and demand centres of gravity, but clearly much closer to the demand side, due to the higher cost of transport used for secondary transport or delivery to the market. The optimised location moves slightly lower in the graph towards the supply centre of gravity. The ideal location for a distribution centre is consequently at this optimised position.

Table 3: Centre-of-gravity calculations.

Location	X (km)	Y (km)	F (ton)	R (R/ton.km)	F×R	X×F×R	Y×F×R	d (km)	Cost (R)	F×R/d	X×F×R/d	Y×F×R/d
Customer 1	38.00	42.00	20.0	R 1.20	24.0	912.0	1008.0	27.76	R 666.20	0.86	32.85	36.31
Customer 2	10.00	45.00	10.0	R 1.20	12.0	120.0	540.0	30.39	R 364.71	0.39	3.95	17.77
Customer 3	22.00	10.00	15.0	R 1.20	18.0	396.0	180.0	8.27	R 148.93	2.18	47.86	21.76
Customer 4	15.00	39.00	5.0	R 1.20	6.0	90.0	234.0	22.82	R 136.95	0.26	3.94	10.25
Customer 5	45.00	11.00	12.0	R 1.20	14.4	648.0	158.4	22.14	R 318.80	0.65	29.27	7.15
Supplier 1	10.00	20.00	11.0	R 0.60	6.6	66.0	132.0	14.14	R 93.35	0.47	4.67	9.33
Supplier 2	38.00	22.00	11.0	R 0.60	6.6	250.8	145.2	14.55	R 96.01	0.45	17.24	9.98
Sea port	20.00	2.00	20.0	R 0.60	12.0	240.0	24.0	16.52	R 198.24	0.73	14.53	1.45
Airport	10.00	10.00	20.0	R 0.60	12.0	120.0	120.0	16.14	R 193.72	0.74	7.43	7.43
Demand CoG	29.11	28.50	62.0					11.65				
Supply CoG	18.19	11.32	62.0					8.87				
Combined CoG	25.47	22.77	124.0					4.97				
Optimised CoG	24.00	18.02						0.00				
Total demand					74.4	2 166.0	2 120.4		R 1 635.59	4.3	117.9	93.2
Total supply					37.2	676.8	421.2		R 581.32	2.4	43.9	28.2
Total combined					111.6	2 842.8	2 541.6		R 2 216.92	6.7	161.7	121.4

Table 4: Centre-of-gravity optimisation iteration.

Optimisation	X (km)	Y (km)	Cost (R)	d (km)
Distance	24.01	18.03	R 2 216.92	0.00
Optimised CoG	24.00	18.02		
0	25.47	22.77	R 2 243.37	1.82
1	25.20	20.97	R 2 227.54	1.09
2	24.88	19.93	R 2 221.51	0.69
3	24.61	19.29	R 2 218.98	0.46
4	24.42	18.87	R 2 217.86	0.30
5	24.28	18.60	R 2 217.35	0.21
6	24.19	18.41	R 2 217.12	0.14
7	24.13	18.29	R 2 217.01	0.10
8	24.09	18.20	R 2 216.96	0.07
9	24.06	18.14	R 2 216.94	0.04
10	24.04	18.10	R 2 216.93	0.03
11	24.03	18.08	R 2 216.92	0.02
12	24.02	18.06	R 2 216.92	0.01
13	24.01	18.04	R 2 216.92	0.01
14	24.01	18.03	R 2 216.92	0.01
15	24.01	18.03	R 2 216.92	0.00

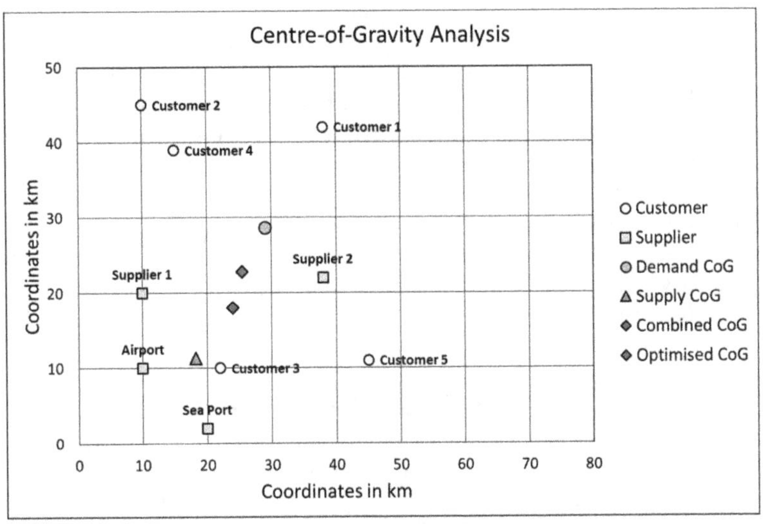

Figure 11: Centre-of-gravity analysis graph.

In conclusion, it is important to mention that the centres of gravity should not be regarded necessarily as the answer. It provides a good starting point on macro level of looking for the best location, but various qualitative factors on micro level such as listed in Table 1 should be considered, as well as the total logistics costs discussed in the next section.

9 TOTAL LOGISTICS COSTS

Once the location of the centres of gravity has been determined and alternative logistics networks configured accordingly, the individual cost components should be calculated for the logistics elements or functions to determine the total logistics costs. Rushton et al. [18] suggest that the following costs should be considered:

- Primary transport cost (inbound; supply from manufacturers or suppliers to distribution centres or other intermediaries);
- Secondary transport cost (outbound; delivery from distribution centres to the market);
- Warehousing cost (including capital and operational costs);
- Inventory holding costs (including capital, service and risk costs); and
- An information system costs (including order processing costs).

Other costs of less importance can include communication costs, protective packaging, direct and indirect taxation, cost of reverse logistics and procurement costs. These costs are usually relatively small but it is important to include those if the amounts are significant and can have a determining impact on the total logistics costs.

The analysis of the total logistics costs recognises that many more costs are incurred through the provision of availability than just transport and warehousing costs. The total least cost might not be where the transport cost is minimum or where inventory or information costs are minimum. Managing transportation, inventories and order processing activities collectively could lead to substantial cost reduction when compared with managing the functions in isolation.

In practice, primary transport, secondary transport, warehousing and inventory holding costs are most important and should be considered. Most of the costs are relatively easily available and where not, industry averages can be used, but in an effort to simplify the process, the analysis should be done by comparing relative rather than absolute numbers. The result of the comparison is the same, but the effort in only considering those cost elements that differ in the different alternatives, rather than a zero-based activity based costing exercise, is significantly less.

The relative relationship between the mentioned costs components are shown in Fig. 12. It is clear that there are significant trade-offs between the different components, such as between secondary transport and warehousing costs. The transport manager will definitely choose decentralisation with multiple depots to be closer to the market and travel less distance

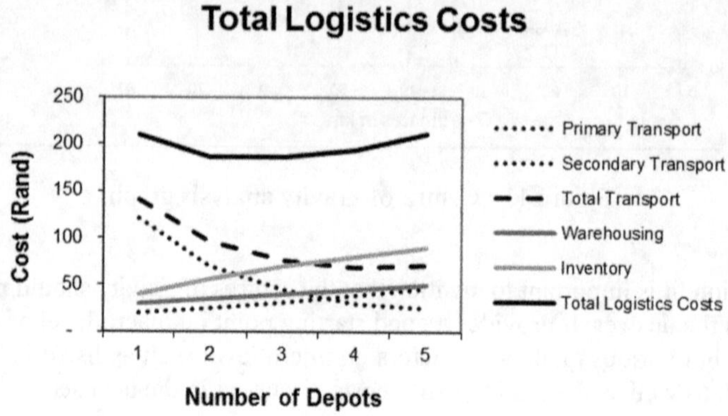

Figure 12: Total logistics costs.

while the warehouses manage will choose centralisation due to obvious cost savings. The answer lies in the total costs and hence the importance to make decision based on total logistics costs and not the individual components.

In conclusion, although the total logistics costs provide a clear quantitative result, other qualitative factors such as service levels, corporate strategy and corporate social responsibility should also be considered before a final choice is made. However, total logistics costs should probably be the most important and relevant performance indicator in the choice of a suitable logistics network.

So, centre-of-gravity analysis provides the answer to where the facilities should be located while total logistics costs provides the answer to how many facilities should be operated. This leads into the last part of the chapter that covers the development of the most appropriate channel strategy.

10 CHANNEL STRATEGIES

Once the ideal logistics network has been identified, based on centre-of-gravity analysis on macro level and total logistics costs have been calculated to assist with micro level comparison, the most appropriate channel strategy should be developed.

There are various approaches to address the development of this strategy, but the supply chain design (or channel strategy) process consists according to Stock and Lambert [19] of the following steps:

1. Establish supply chain objectives.
2. Formulate a supply chain strategy.
3. Determine supply chain structure alternatives.
4. Evaluate supply chain structure alternatives.
5. Select supply chain structure.
6. Determine alternatives for individual supply chain members.
7. Evaluate and select individual supply chain members.
8. Measure and evaluate supply chain performance.
9. Evaluate supply chain alternatives when performance objectives are not met or when attractive new options become available.

The second step suggests that a supply chain strategy be formulated and Perez [20] submits that supply chain strategies generally conform to one of six types as indicated in Table 5.

This is similar to work done by Gattorna [21] over many years on the development of channel strategies. He is of the opinion that there exist the following generic supply chain types:

- Continuous replenishment supply chain: Very predictable demand from known customers; easily managed through tight collaboration with these collaborative customers; focus on retention of customer relationships;
- Lean supply chain: Regular pattern of demand; quite predictable and forecastable although may be seasonable; tend to be mature low risk products/services; focus on efficiency;
- Agile supply chain: Usually unplanned at least until the last possible moment. May result from promotions; new product launches; fashion marketing; unplanned stock-outs; or unforeseen opportunities. Focus on the service-cost equation; and
- Fully flexible supply chain: Unplanned and unplannable demand due to unknown customers with exceptional, sometimes emergency, requests. Focus on providing creative solutions at a premium price.

Table 5: Generic supply chain models [20].

Orientation	Supply chain model	Description
Efficiency	Continuous-flow	A supply chain focused on building synergies supported by collaborative relationships. Low working capital for customers is the main deliverable. Workload is smoothed by customer's demand. Order penetration point is "make to stock" in order to ensure medium-high utilisation rates with a high level of perfect orders. Production, driven by market demand, is designed to replenish predefined inventory levels.
	Efficient	A supply chain focused on lowest cost. Used for functional and commoditised products. Weakest value proposal when used in an isolated way, useful when used as a backbone for supporting mixed strategies. Efficiency is supported by a high asset utilisation rate based on "make to forecast" model in order to maintain continuous production and ensure an optimal production sequence, thereby reducing setup time.
	Fast	A supply chain focused on competitive cost and continuous portfolio renewal. Short time from "idea to market" and an affordable cost are main deliverables. Used for functional and short-lifecycle products oriented to fashion. Demand is "pushed" by a collection forecast. Fast product development is critical capability. Applied by "fashion creators", especially in industries such as apparel and beauty products.
Responsiveness	Custom-configured	A supply chain focused on product configurability, where several product features are customised among a set of limited options. Normally oriented towards the end user. Workload before product's decoupling point is smoothed by forecast. Extra capacities after decoupling point and design for easy assembly or postponement are key elements for agility. Order accuracy is a relevant factor in order to ensure fulfilment of customised orders.
	Agile	A supply chain focused on responsiveness to unpredictable demand. Used for "exclusive" and short-lifecycle products. Asset utilisation is not sufficiently relevant in the total cost. Workload has medium-sized peaks and valleys. Customised/exclusive products and short order cycle are main deliverables. Extra capacity and common components/materials are key elements for agility. Normally oriented towards industrial customers.
	Flexible	A supply chain focused on solving problems, sometimes including customer's emergency situations, or on proposing solutions, perhaps at the level of engineering projects. Mainly used for products with high level or added services. Typically price is irrelevant for customers due to uniqueness of the solution. Capacity on standby, capacity pooling, and reconfigurable processes are key factors for ensuring flexibility. Relationships with customers tend to be shorter-term.

This last type is further split between a business event strategy in an entrepreneurial environment and a humanitarian response strategy in an emergency environment.

Management of the strategies requires different approaches to activities and challenges in the supply chains in general and network design in particular. Gattorna [22] provides a comparative analysis of the different strategies and his opinion on the differences in channels of distribution (or network design), is indicated in Table 6.

The generic supply chain strategies and associated logistics networks all contribute to the objective of developing and implementing a supply chain that will be able to compete effectively on cost, service levels and whatever other performance indicators require. The solution is not 'one size fits all' but a process to develop a customised supply chain with support of modelling techniques such as centre-of-gravity analysis.

Table 6: Supply chain strategies and network design [22].

Supply chain strategy	**Channels of distribution (Network design)**
Continuous replenishment supply chain	Either direct or via trusted outlets
Lean supply chain	Wide distribution through multiple channels for maximum accessibility
Agile supply chain	Fewer, more direct channels to access consumers
Fully flexible supply chain – business event	Limited and very targeted
Fully flexible supply chain – humanitarian	As many as needed in a given situation

11 CONCLUSION

The purpose of this chapter was to explore the importance of heuristic modelling in the development of a suitable logistics network and most appropriate channel strategy. The need for and application of the centre-of-gravity analysis technique was presented and discussed with a practical example.

It is clear that there is not necessarily one best solution but an integrated system of nodes and links (facilities and channels) in the logistics network that provides for supply chains to compete effectively. The most appropriate channel strategy is highly likely product specific and developed for specific conditions in a specific supply chain.

REFERENCES

[1] Christopher, M., *Logistics & Supply Chain Management*, Pearson Education Limited: Harlow, pp. 13–15, 2011.

[2] CSIR, *The 10th Annual State of Logistics Survey for South Africa – 2013*, Pretoria, pp. 7–13, 2014.

[3] CSCMP, 25th Annual State of Logistics Report, Lombard, p. 2, 2014.

[4] De Villiers, G., Nieman, G. & Niemann, W., *Strategic Logistics Management – A Supply Chain Management Approach*, Van Schaik Publishers: Pretoria, pp. 213–214, 2008.

[5] Christopher, M., *Logistics & Supply Chain Management*, Pearson Education Limited, Harlow, pp. 265–266, 2011.

[6] Stock, J.R. & Lambert, D.M., *Strategic Logistics Management*, McGraw-Hill: New York, pp. 702–704, 2001.

[7] Rushton, A., Croucher, P. & Baker, P., *The Handbook of Logistics & Distribution Management*, Replika Press: India, pp. 50–58, 2010.

[8] Lambert, D.M. & Stock, J.R., *Strategic Logistics Management*, Irwin: Boston, pp. 75–77, 1993.

[9] De Villiers, G., Fourth-party logistics, *Logistics News*, April 2007, p. 15, 2007.

[10] Gattorna, J.L., *Strategic Supply Chain Alignment – Best Practice in Supply Chain Management*, Gower Publishing Company: Aldershot, pp. 425–433, 1998.

[11] Stock, J.R. & Lambert, D.M., *Strategic Logistics Management*, McGraw-Hill: New York, pp. 410–417, 2001.

[12] Ackerman, K., *Warehousing Profitably*, Ackerman Publications: Columbus, pp. 265–267, 2011.

[13] Simchi-Levi, D., Kaminsky, P. & Simchi-Levi, E., *Managing the Supply Chain – The Definitive Guide for the Business Professional*, McGraw-Hill, pp. 71–80, 2004.

[14] Blanchard, D., *Supply Chain Management – Best Practices*, John Wiley & Sons: New Jersey, pp. 105–108, 2010.

[15] Simchi-Levi, D., Kaminsky, P. & Simchi-Levi, E., *Designing and Managing the Supply Chain – Concepts, Strategies, and Case Studies*, McGraw-Hill, pp. 112–117, 2000.

[16] Coyle, J.J., Langley, C.J., Gibson, B.J., Novack, R.A. & Bardi, E.J., *Supply Chain Management – A Logistics Perspective*, South-Western Cengage Learning: Mason, pp. 542–549, 2008.

[17] Bowersox, D.J. & Closs. D.J., *Logistical Management – The Integrated Supply Chain Process*, McGraw-Hill: Singapore, pp. 554–561, 1996.

[18] Rushton, A., Croucher, P. & Baker, P., *The Handbook of Logistics & Distribution Management*, Replika Press: India, pp. 120–127, 2010.

[19] Stock, J.R. & Lambert, D.M., *Strategic Logistics Management*, McGraw-Hill: New York, pp. 79–80, 2001.

[20] Perez, H.D., Supply chain strategies: Which one hits the mark?, *CSCMP's Supply Chain Quarterly*, Special Issue 2014, pp. 30–40, 2014.

[21] Gattorna, J.L., *Dynamic Supply Chains – Delivering Value Through People*, Prentice Hall: UK, pp. 51–56, 2010.

[22] Gattorna, J.L., *Dynamic Supply Chains – Delivering Value Through People*, Prentice Hall: UK, pp. 79–86, 2010.

INTEGRATED LAND-USE AND TRANSPORTATION MODELLING IN DEVELOPING COUNTRIES: USING OPENTRIPPLANNER TO DETERMINE LOWEST-COST COMMUTE TRIPS

LOUIS WALDECK & QUINTIN VAN HEERDEN
Spatial Planning and Systems, Built Environment, Council for Scientific and Industrial Research, Pretoria, South Africa

ABSTRACT

Large cities are amongst the most complex production systems ever built. In this chapter the role of Land Use Transportation Interaction (LUTI) models in supporting the equally complex urban planning processes that shape our cities is being explored. Specific reference is made to developing countries where poverty, income disparities, and mounting demands for basic services pose even greater challenges to urban planners. The paper refers to the urban simulation model that was developed for South Africa by adapting UrbanSim and OpenTripPlanner and how this combination allowed practitioners to reduce the onerous data preparation and simulation times that have become characteristic of leading micro-simulation LUTI models. In the chapter the case for using the monetary cost of commuting as an acceptable proxy for the generalised cost because low-income households in developing countries tend to be more sensitive to the cost of a trip than the duration is demonstrated. The results from two case studies in different South African cities are presented and the combined model is validated by simulating a ten-year period between the last two population censuses.

Keywords: transportation modelling, urban modelling, integrated land-use and transportation, developing countries, UrbanSim, OpenTripPlanner.

1 INTRODUCTION

In the developing world, mobility challenges and the role of government in the provision of infrastructure are significantly different from those in the developed world. Confronted by extreme poverty, income disparities and mounting demands for basic services such as clean water, sanitation and health care, urban planners in developing countries face daunting challenges, Cervero [1]. In Sub Sahara Africa, 62 percent of urban households reside in shantytowns and 64 percent of those with formal jobs lived on less than US$1.25 per day in 2010, United Nations [2]. In South Africa, 20 percent of households lived in informal settlements (shantytowns) and a further 11 percent in informal backyard dwellings in 2011 according to Statistics South Africa [3]. The median personal income was about US$14 per day based on average exchange rate of US$1 = ZAR6.9 in 2011.

Another challenge to the Government of South Africa is striking a balance between investment in 'economic infrastructure' such as transportation and energy supply systems and 'social infrastructure and services' including water, sanitation, health and a variety of grants/subsidies aimed at alleviating the plight of the poor without expecting a monetary return on the investment. Public transportation has attracted massive investments since the promulgation of the National Land Transportation Act of 2009. Like in most cities around the world the new public transportation services are subsidised, but probably more so as a service to the poor and this chapter will endeavour to highlight the significant benefits of subsidisation. The jury is however out on whether target densities will be reached within the designated transportation corridors to achieve the ridership levels required to contain subsidies.

Since urban decision making processes can be adversarial due to differences in opinion about the policy scenarios that should be adapted to build a more sustainable city, it is no wonder that a myriad of planning support models have over the years made their appearance to assist decision makers by simulating the likely outcome of policy scenarios. The relative success of a policy scenario in achieving one or more objectives is then evaluated by comparing a number of predefined indicators. For example, in a public transport corridor densification scenario, the success of public transport and higher density zoning could be measured by comparing a simple indicator such as the resultant population density within the confines of the corridor, with and without public transport.

The value of these models is not in predicting the future, which it obviously cannot do, but in providing participants in the planning process with a better understanding of cause and effect between different components of the urban system, which ranks among the most complex production systems ever built and in discovering common ground between different departments and interest groups that could lead to compromise. We thus argued that such tools are even more relevant in developing countries due to the additional challenges outlined at the outset.

The class of models suitable for this type of work is commonly referred to as Land Use Transportation Interaction (LUTI) models. The literature abounds with a daunting number of LUTI models and many papers have been written on the topic of comparing models in terms of criteria such comprehensiveness, theoretical foundations, data requirements, etc. In one of the earlier papers Wegener [4] concluded that "*The most promising technique for activity-based land use and transport modelling is micro simulation which makes it possible to reproduce the complex spatial behaviour of individuals on a one-to-one basis*". Nearly a decade later Wegener [5] concluded that the trend towards greater disaggregation has gone mainstream, activity-based travel models have become the state-of-the-art and agent-based land-use models are proliferating. Simulation runs for leading micro-simulation models are regrettably now measured in terms of days or weeks, not hours.

In South Africa the Council for Scientific and Industrial Research (CSIR) developed an Urban Simulation Model that has been used to simulate spatial growth patterns 30 years into the future, to better understand the future demand for infrastructure, facilities and services including water, sanitation, electricity and transportation. The model was initially based on UrbanSim and MATSim but is currently based on adapted versions of UrbanSim and OpenTripPlanner (OTP). The change from MATSim to OTP was partly motivated by the need to achieve a significant improvement in computational efficiency and partly to build a model that would be more representative of a developing country like South Africa.

In this chapter the rationale for adapting OTP is presented and applied as a substitute for MATSim, the nature of the interaction between UrbanSim and OTP and discuss the advantages and limitations of the approach. The results of successfully applying the Urban Simulation Model in the Cities of Tshwane and Nelson Mandela Bay is being presented. To start off with, the scope of a typical LUTI model is explored before providing a brief overview of UrbanSim.

2 SCOPE OF LAND USE TRANSPORTATION INTERACTION MODELS

In a comparative study of twenty contemporary models Wegener [4] identified the eight important urban subsystems in Fig. 1, of which UrbanSim addresses 7. Other models addressed a minimum of two and maximum of eight subsystems.

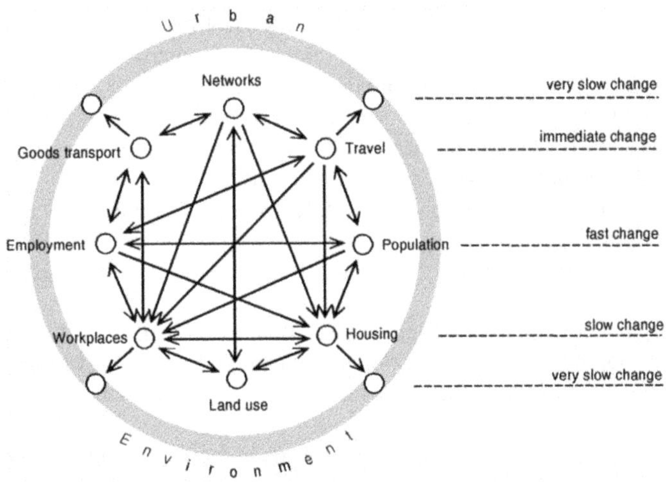

Figure 1: Web of urban processes (from Waddell [6], after Wegener [7]).

From the perspective of the actors involved in various subsystems one could summarise the above diagram as follows: Households decide where they live, where they work and how they commute between the two. Employment opportunities are created by the location decisions of firms. Developers provide housing and floor space for conducting business after calculating the expected return on investment of development proposals and constructing proposals that maximise their profit. Government provides transportation networks and regulates physical development by virtue of the rights associated with each parcel of land within a policy environment.

The spatial distribution of residential, business and industrial land uses in particular determines the location of human activities such as living, working and shopping. The demand for transportation is a direct result of the separation of these activities. The supply of transportation in turn determines how accessible places of interest are, which may in turn influence the choice of location of households and thereby the demand for transportation to complete the 'land-use transport feedback cycle'.

The literature abounds with a daunting number of available LUTI models. A decade ago already Wegener [4] pointed out that there were significant variations among models with respect to comprehensiveness, model structure, theoretical foundations, modelling techniques, dynamics, data requirements and calibration and validation. Based on his more recent research findings from the EU-funded SustainCity programme, Wegener [5] found that the primary difference between models can be reduced to: (1) whether a model represents the urban system as a dynamically changing system or one in equilibrium, and; (2) the degree of disaggregation allowed by a model to capture the heterogeneity in human behaviour resulting from individualisation in modern societies. The latter is particularly important due to various forms of inequality that are characteristic of the South African context.

In another review by Simmonds et al. [8] UrbanSim was considered as best practice (with two other LUT models: DELTA and IRPUD) in terms of quasi-dynamic models (there are no fully dynamic models available yet) by virtue of subsystem-specific sub models.

From the foregoing it can be concluded that UrbanSim finds itself in an elite group of state of the art models. In the next section we provide a brief overview of UrbanSim before returning to OpenTripPlanner.

3 OVERVIEW OF URBANSIM

UrbanSim recognises the property market as the central construct of the urban system and accounts for the current stock of real estate (and its spatial distribution) as well as the supply and demand of new stock. In Fig. 2, households and businesses represent the consumers of residential and non-residential stock respectively. The choice of location by these agents is accounted for by the occupation of space in a building.

Developer agents, representing the supply side, use occupancy rates as a feedback signal to determine the rate at which new stock can be built and sold. Supply and demand are mediated by price, another feedback signal which influences the return on investment calculation done by developers.

Governments set policies that either regulate the use of land or influence development through pricing policies such as developer contribution fees. Governments also build infrastructure and provide social facilities such as libraries, clinics, community halls, parks, sports field and fire stations.

Transportation infrastructure and the quality of social facilities influence the attractiveness of locations for different consumers.

Households have particular characteristics that influence their preferences for housing of different types at different locations. Similarly, businesses have preferences that vary by industry and number of employees for alternative building types and locations. Two models are prominent in modelling these choices, the Household Location Choice Model and the Employment Location Choice Model. Both are discrete choice models based on the pioneering work of McFadden on Random Utility Maximization Theory (McFadden [10] and McFadden [11]). This approach determines the probability of a choice from a set of available alternatives based on the characteristics of the chooser and the attributes of the alternatives, depending on the relative utility that an alternative offers the chooser.

The systematic component of a utility (as opposed to the random component) is expressed as a linear combination of estimable coefficients multiplied by independent alternative-specific variables that may be interacted with the characteristics of the household making the choice. The coefficients are estimated using maximum likelihood methods built into UrbanSim. For choice models the estimation can be based on the stated or observed behaviour of agents. All research by the authors so far has been based on observed behaviour derived from sources such as census and household travel surveys.

Examples of alternative-specific variables often used in utility calculations include the distance, travel time and generalised cost of commuting. The term generalised cost refers to

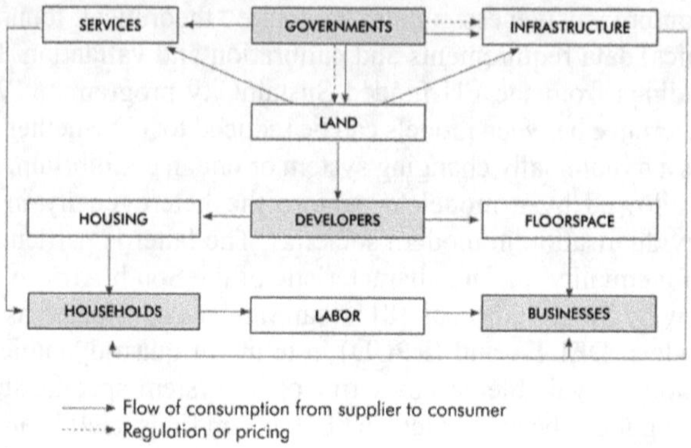

Figure 2: Interaction of agents with urban markets (from Waddell [9]).

the sum of the monetary cost and the value of time spent undertaking the trip, which varies according to the traveller's income and the purpose of the trip.

The choice of variables is more of an art than a science. The process usually starts by considering some variables derived from urban economic theory. Guided by various indicators which estimate the relative importance and statistical significance of each variable, variables are added or deleted to maximise the overall explanatory power of the model. In South Africa the process presented a few more challenges, such as a negative correlation between low-income households and proximity to work opportunities. Whereas this would defy urban economic theory in a developed country, it is still a reality in South Africa due to the politically motivated spatial planning policies of the past.

It is unable to describe all the adaptations that were made to both UrbanSim and OpenTripPlanner in one chapter. The focus will thus only be on OTP and UrbanSim will be dealt with in future publications.

4 RATIONALE FOR USING OTP AS SUBSTITUTE FOR MATSIM

The investigation into using OTP as a substitute for MATSim was prompted by the research of Nicolai and Nagel [12] to determine if measures of accessibility based on pairs of locations could be replaced by measures based on attributes of the location itself, since this would significantly improve computational performance. As such, the study by Nicolai and Nagel was undertaken with the objective of improving the interaction between UrbanSim and MATSim rather than finding ways of avoiding it. As long as travel times or generalised costs are used in the calculation of accessibilities, such calculations will have to be based on the congested state of transportation networks.

But what if the monetary cost of a commute trip can be used as acceptable proxy for generalised costs in South Africa, and perhaps other developing countries? In that case, cost and distance-based measures of accessibility would no longer depend on the congested state of networks and since the monetary costs and network-distances can be obtained from OTP, no need for interaction with MATSim, at least not for the purpose simulating urban growth, exists.

The remainder of this section explains how this possibility was investigated. Firstly, the underlying causes of long simulation runs were examined, then a high resolution accessibility measures as a means of improving computational efficiency is introduced and finally the argument, based on the household income distribution of South Africa that a model based on monetary costs could be a more representative model of reality in South Africa is developed. From this, the potential for other developing countries is deduced.

4.1 Interaction between UrbanSim and MATSim

Travel times and generalised costs are usually obtained from a series of interactions between UrbanSim and a transportation model that represents the congested state of the transportation network, such as EMME, VISUM or MATSim. Interaction is achieved by way of an exchange of data at the start of every UrbanSim simulation year as illustrated in Fig. 3. The effect of this simple action is unfortunately to increase the total execution time for a 30-year simulation period by many hours or even days depending on the size of the study area. Although some progress has been made with parallelising UrbanSim (Awaludin and Chen [13]) and MATSim (Raney and Nagel [14]) it is unlikely that the coupled system will benefit in the same way from advances in parallel computing because one model has to finish before the other can start with the next year. It is therefore common for modellers to introduce austerity measures such as restricting interaction to 5 or 10 year intervals.

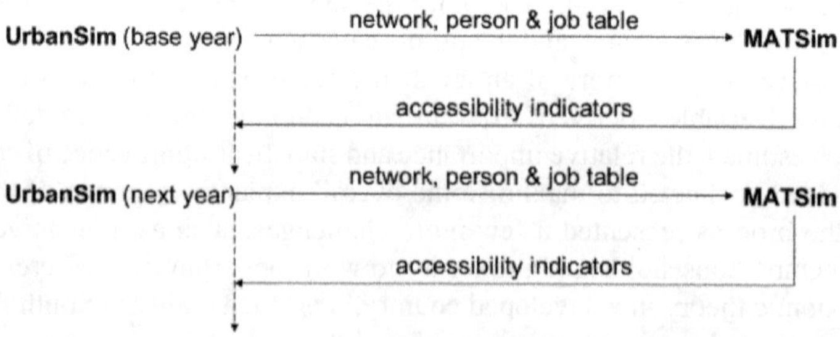

Figure 3: Sequence of interaction between UrbanSim and MATSim (from Nicolai and Nagel [12]).

4.2 High resolution accessibility measures

In an attempt to resolve some of the above constraints, Nicolai and Nagel [12] investigated the notion of high resolution accessibility to determine if measures of accessibility based on pairs of locations could be replaced by measures based on attributes of the location itself. The study was prompted by the observation that many of the variables used in specifications of the Household Location Choice Model all over the world appeared to be simple accessibility measures anyway, for example distance to workplace, jobs within 30 min drive time, etc. This not only agrees with the experience of the authors and that of other papers from the EU-funded Sustain City project, for example Zöllig Renner et al. [15] but appears to also be a fair representation of household decision-making processes. Few, if any, households consult transportation models when considering the location of a new house. For road trips they probably assume that most roads are congested during peak times and use distance as proxy for comparing the expected commute time from the new location to their current commute time.

Hansen [16] defines accessibility as the potential of opportunities for interaction. If locations are otherwise equal, a location with easier access to activities like work, leisure or shopping is more attractive than locations with less access and have greater potential for residential development. Moeckel [17] confirms that this approach is also true for businesses.

In a study, Nicolai and Nagel defined the accessibility of location i as log sum of generalised cost C_{ij} over all locations j:

$$A_i := ln \sum_j e^{-\beta C_{ij}}. \quad (1)$$

In high-resolution accessibility calculations, there are two resolutions to consider: one that determines for how many origins i the accessibility is to be computed and another that determines to what level the destinations j need to be resolved.

The number of origins was made variable in the form of a grid measuring 250, 500, 1000, 2500, 5000, 7500 and 10000 meter per side. The number of destinations is taken as a 10% sample of all job locations (more precisely the network nodes closest to the buildings with these jobs).

Nicolai and Nagel applies an existing UrbanSim/MATSim model of the Puget Sound Regional Council (which includes Seattle) to investigate the feasibility of calculating high resolution accessibilities for each of the seven origin square sizes. A generalised costs C_{ij} from the first MATSim year is used as representative of the congested state of the network.

The results demonstrate that it is computationally feasible to, for every origin; calculate a weighted sum over all possible destinations separately, rather than aggregating them into zones. This removes all zonal artefacts from the destination side of the computation, such as job opportunities in neighbouring zones being easier to reach than opportunities within the zone itself. It is also concluded that the spatial resolution of the origins had a strong impact, with resolutions smaller than 1000m x 1000m producing diminishing returns.

4.2.1 Origins and destinations

Regarding the spatial resolution of origins and destinations, UrbanSim applies three geographies: user-defined grid, land parcels or user-defined zones. The research focus of this chapter is, thus far, based on something between parcels and zones. It commences with individual property boundaries but later on combines adjacent parcels in a street block that are indistinguishable in terms of a number of attributes into a single parcel and represent all the (similar) buildings on the individual parcels as one pseudo building per street block. This reduces the number of parcels and buildings by a factor of roughly 20 with distinct advantages for GIS intensive data preparation and some reduction of UrbanSim execution times, without affecting the explanatory power of the model. Households residing in the same street block are not combined and dissimilar properties such as residential and commercial are likewise not combined.

With OTP it is perfectly feasible to determine an accessibility indicator for every street block even if this requires the lowest cost route over all modes of transport to be calculated from every street block to every other street block. This is made possible by the fact that OTP 'remembers' the result of previously considered routes to avoid unnecessary recalculation.

Mindful of the conclusion reached by Nicolai and Nagel [12] that little is gained by using origins smaller than about 1 km^2, street blocks that are typically 20 to 50 times smaller, is considered to be wasteful. Another factor that influences the choice of resolution is that the notion of a street block is only useful for areas that are already developed or in the process of being established. In outlying areas the smallest property boundary in the base year of a simulation could be as large as a farm. Using such parcels as unit of analysis in longitudinal studies of urban growth could lead to misleading results due the Modifiable Areal Unit Problem (MAUP). To avoid such complications, it was decided to use 'modified small areas' or 'modified sub places' both as origins and destinations for OTP and as unit of analysis for UrbanSim. In the rest of the chapter these will simply be referred to as 'zones'. 'Sub places' and 'small areas' are two of the units of analysis at which census information is released by Statistics South Africa (StatsSA). The authors modify these areas by retaining the smaller ones in built-up areas, sometimes combining very small ones and subdividing the larger ones (and where necessary the corresponding parcels) in order to obtain a more even size distribution. In the case of Nelson Mandela Bay the zones on modified small places with a median area of 0.9 km^2 and a maximum of 2.3 km^2 are being used.

Note that while the number and size of zones affect the number of lowest cost routes to be found by OTP, it does not affect the street-block resolution of UrbanSim simulations, merely the spatial resolution to which the results are aggregated afterwards for analysis and interpretation purposes.

4.2.2 Changing location of simulated jobs

One of the initial attractions to investigate the notion of accessibility measures based on the monetary cost of commuting is that such measures no longer depends on the congested state of networks and only needs recalculation when there is a significant change to the road or transit network.

If the accessibility measures are passed back to UrbanSim in the same way as contemplated in Fig. 3, it is only be valid for one year because the location of jobs keep changing from one simulation year to the next. To avoid this it was decided to store the OTP results of the lowest-cost routing for different modes of transport per Origin–Destination (OD) pair. This allows the use standard UrbanSim functionality for defining variables to calculate measures of accessibility based on the correct job locations for each simulation year. It also opens up numerous other possibilities to customise accessibility to jobs in different sectors of the economy, for households of different income groups, etc.

This solution opposes the efforts of Nicolai and Nagel [12] to get rid of storing OD pairs, but is well worth it if it avoids repeatedly running OTP or MATSim. For example, in Nelson Mandela Bay with 2672 zones, the OD matrix has 7.14 million rows and 6 columns, taking up 512 MB of storage. All OTP related processing take about 30 minutes on an ordinary laptop computer and as mentioned before it only has to be repeated when a different scenario requires the network to change.

If the OD matrix becomes a constraint for large study areas, the size could be cut in half by changing it to a symmetrical matrix on account of the fact that the cost of a trip from A to B will only differ marginally from the trip from B to A where one-way streets are involved. For the focus of this research, it does not pose a limitation.

4.3 Accessibilities based on the monetary cost of commuting

Having explain how measures of accessibility such as the log sum defined in eqn (1) may be applied as an UrbanSim variable and used in the Household Location Choice Model to influence the location choices of households, it is now possible to move considering whether accessibility measures based on the generalised cost of commuting can be replaced by measures based on monetary costs in South Africa and perhaps elsewhere in the developing world. This premise does not imply that the same approach is to be followed in the developed world.

The difference between using monetary costs and generalised costs is that the latter depends on the congested state of the network and introduces another complication in the form of the value of time, which is not easy quantifiable, even in the developed world. One of the difficulties with the value of time is the difference between the stated and observed behaviour of households. If high-income households residing in security areas such as golf and eco estates on the periphery of South African cities, are surveyed, they will put a high value on time. The conclusion may thus be drawn that location choice adds to daily travel time that location choices travel time is of less importance and less valued. The answer is often that there are other factors which are even more important than travel time, such as safety and security considerations, which are often quoted as a reason for the popularity of such estates.

Apart from the value of time, the argument for using measures of accessibility based on monetary cost in South Africa is based mainly on an income distribution which has a large proportion of poor households. It is well known that low-income households have to spend proportionally more on transportation than higher income households and it is concluded from the research that the lowest income groups (typically found in informal settlements or backyard dwellings) follow scarce job opportunities to minimize expenditure on transportation, often by opting for non-motorised modes (especially walking) of transport.

4.3.1 Household income

For the purpose of determining what percentage of households could be considered to belong to 'lower income groups' various studies dealing with the literal and other more nuanced definitions of the 'middle class' were consulted, for example Visagie [18] and Visagie and Posel [19]. The authors of this chapter eventually decided to define the 'lower income group' as all households with income less than the lower limit of the more nuanced 'middle class' shown as 'Relatively Affluent Middle' in Fig. 4 below. This point of departure is based on the National Income Dynamics Study (NIDS) of 2008. Although the actual breakpoints differ slightly from the Census of 2011, the overall conclusion is that about 65% of households are classified in the category of 'low income'.

Since the ratio of household expenditure on a given mode of transportation to household income increases with decreasing income, it would be expected that the location choices of poorer households to be more sensitive to trip cost than travel time. This is one reason why accessibility measures based on monetary costs are a more representative model for SA than one based on travel times and generalised cost.

It, however, should not be confused with the determinants of modal choice and location preference. Due to this statement, it is useful to explore the results of the National Household Travel Survey of 2013 [20].

4.3.2 National Household Travel Survey

The figures quoted below differ from the published Statistical Release [20] because they are based on an analysis of a subset of data representing the six largest metropolitan cities that the Urban Simulation Model would typically be applied to.

In application of research related to travel the following findings are applicable: In the bottom four quintiles 5% of workers walk to work; another 80% use public transport (i.e. minibus taxi, bus, and train) and a mere 15% private transport as driver or passenger. In the top quintile the corresponding figures are 2% of workers walk, 35% use public transport

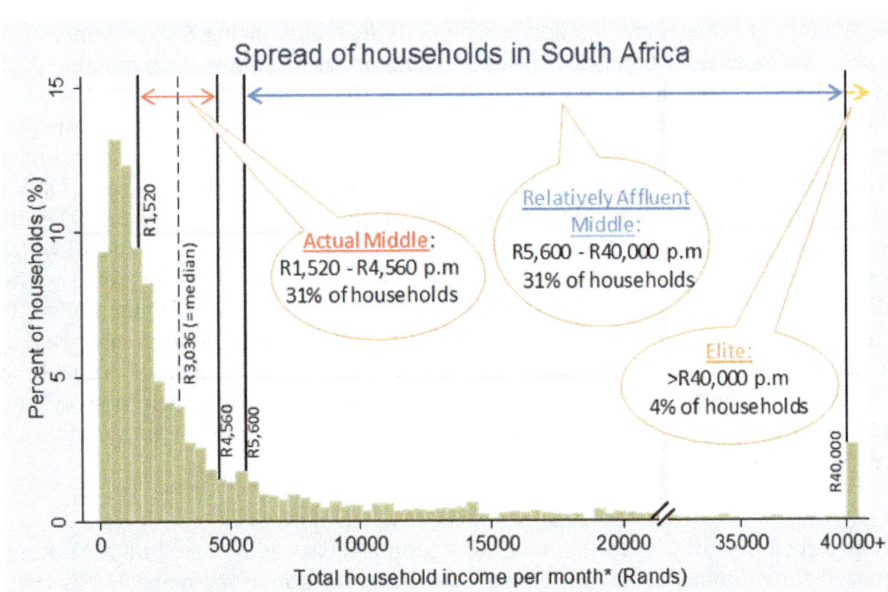

Figure 4: Household income distribution (from Visagie [18]).

and the remaining 63% private transport. Regarding the factors that influence the preferred choice of a particular mode the following finding applies: In the bottom two quintiles more workers cited travel cost than travel time, roughly equal numbers in the third quintile and more workers cited travel time than travel cost in the top two quintiles. Also of interest is that factors such as flexibility of time and destination increase from 5% in the bottom four quintiles to 15% in the upper quintile.

The interpretation of the abovementioned results is as follows: For 60% to 80% of households, cost is critical. Based on the observation that the percentage of workers that walk all the way to work peaks in the bottom three quintiles it is unlikely to be motivated by healthier living considerations, more likely a survival or household budget balancing mechanism. Even where travel time is cited as more important than travel cost it is probably expressing a preference for a faster minibus taxi service. Of the 80% of workers that rely on public transportation 51% use minibus taxis, nearly double the number of workers that rely on train (15%) and bus (13%) services. Due to government subsidies, the fare structures for public transport are such that Metrorail fares are the cheapest, bus fares approximately double that, and minibus taxis, which are regarded as public transport but not subsidised, approximately double the bus fares. This creates an environment where commuters from the bottom two quintiles choose Metrorail where available, then bus and then minibus taxis. For commuters from households in the next two quintiles this may change to Metrorail when and where convenient, then a choice between a slower bus service and a faster minibus taxi or BRT service depending on whether travel time or travel cost are most important to the commuter.

For 20% of households in the upper quintile, 63% of workers use private transport. These households have the means to make lifestyle choices and some prefer to live in security, golf or eco estates that are not centrally located. It may be expected that travel time would be important to these households but as pointed out earlier this may be overridden by safety and security considerations or there may simply be no public transport available in such areas.

In conclusion, the research shows that a model based on monetary costs offers an acceptable if not better representation/reflection of the reality facing 80% of households in South Africa. Does this approach sacrifice anything in terms of how the remaining 20% of households are accommodated in terms of the model as developed? To answer this question it must be pointed out that the implementation of UrbanSim deals with five income groups in different sub models of the Household Location Choice Model, allowing completely different variables to be used as predictors of behaviour for each income group. Only two variables, travel time and generalised cost are not available as a result of using OTP. It has been pointed out that it is no trivial matter to quantify the value of time. Even when travel times were available in the past, the research rarely found travel times to be more significant predictors of behaviours than distance-based variables. This is in line with the finding from at least one of the case studies of the SustainCity project, in which Zöllig Renner et al. [15] concluded that the distance to workplace to be the 2nd most important variable with accessibility measures based on generalised cost 4th and 5th most important. Also of interest is that affordability of rent or bond payment featured as the 3rd most important variable, which correlates with the situation in South Africa.

5 MODELLING TRIP COST USING OTP

Having argued that utilities for household location choices can be based on accessibilities calculated from distance and monetary cost variables, the line of reasoning in this chapter, will focus on how to use OTP for such calculations.

5.1 Software

The investigation for the focus of this research commences by considering various software solutions to compute the lowest cost route between a potentially large number of origins and destinations. The routing algorithm was required to consider all modes of transport, including distance limited walking and cycling, private vehicles, unscheduled passenger vehicles and mass transit (bus and rail), as well as possible transfers between any of these modes.

After considering various alternatives, including ArcGIS Network Analyst, NetworkX and OpenTripPlanner it was concluded to select the OpenTripPlanner (OTP). The decision was based primarily on the observation that OTP is purpose-made for routing, whereas ArcGIS and NetworkX are more general purpose tools that would require greater customisation than OTP to achieve the objective stated above. OTP Analyst already makes the transit system model and optimization logic originally developed for point-to-point searches available for one-to-many and many-to-many queries in large batches.

Using OTP for the purpose envisaged in this section involves preparing XML configuration files and then running the JAVA classes for OTP *Graph Builder* and OTP *Batch Processor*. The configuration files are required to set options and point the software to input and output folders and other simulation specific variables such as the maximum walking distance to a public transport stop. The *Graph Builder* converts an OSM file for the study area and a collection of General Transit Feed Specification (GTFS) [21] files into a routable graph object required by the *Batch Processor* to determine the lowest cost route between any number of origins and destinations. The origins and destinations could be provided as GIS point or polygon features or as coordinates in a text file.

5.2 The General Transit Feed Specification

The General Transit Feed Specification (GTFS) is a format developed by Google to publish spatial and temporal information of public transportation schedules. When an agency publishes its schedules in this format it becomes a GTFS feed. The schema allows application developers and transport planners to use the information and build applications like the TriMet Map Trip Planner [22] on top of it. Furthermore, when a GTFS feed is made publicly available on the GTFS Data Exchange website [23], the feed can be used for routing in Google Maps.

A GTFS feed consists of various files, containing detail of the movement and location of vehicles and stations operated by an agency. The minimum files required for a GTFS feed to work are: *Agency, Stops, Routes, Trips, Stop Times, and Calendar*.

The *Agency* table contains information about each transit operator. The *Calendar* table describes the different services that an *Agency* operates and specifies the time periods during which services are available. The *Routes* table lists all the routes operated by an *Agency*. Other fields in the table include the name and type of route. The *Trips* table defines the trips that constitute the routes in the *Routes* table. Routes may be linked to more than one trip. The *Trips* table furthermore contains information about the weekly schedule followed by the *Agency* as well as a direction. The direction allows agencies to distinguish between two trips that operate on the same route, but in different directions. The *Stops* table contains information about the locations where vehicles can be expected to stop, including a name and zone number, which can be used for calculating zone-based fares. The *Stop Times* table describes the sequence of stops followed on each trip in the *Trip* table. The arrival and departure times describe the time of arrival and departure of the vehicle at each stop on the route. The sequence of stops is described with a sequential entry for each stop. The

Frequencies table describes the frequency with which vehicles depart from the first location on the route.

The *Fare Rules* table describes the conditions that need to be satisfied for a fare to be applicable on a specific route. There are different options to specify the rule. One option may be to base the fare on origin and destination zones, which will be applicable to any trips that start and end in these zones. Another option is based on a route, which only then applies to trips that are part of the specific route. Finally, it can specify a specific stop and the fare will only apply to routes containing this stop.

5.3 OTP Workflow

OpenTripPlanner is geared towards providing the fastest or shortest route between one or more origins and destinations. To obtain the lowest (monetary) cost route, the research necessitates the introduction of a number of modifications.

The main process of obtaining the shortest path between two points involves running two classes in OTP, namely *Graph Builder* and the *Batch Analyst*. To run *Graph Builder* it requires a configuration file, an OSM network, and the GTFS feed of public transport agencies to be used in the model. The *Graph Builder* then constructs a graph, which consists of nodes that are connected with edges. Each edge represents a path that can be followed between nodes. Edges can therefore consist of paved road segments, a specific segment on a transit route, or a section between a house and a road, which provides access from the house to the road.

The *Batch Analyst* then uses the graph object as input as well as a configuration file, the GTFS feeds and the origins and destinations between which routes are to be found. For each origin and destination pair, a path is found in the graph. The path consists of a series of edges, which describes the legs of the itinerary (compare the content of Fig. 5)

For the purposes of the research, the *Batch Analyst* in Version 0.10.0 of OTP was adapted as follows: After the *Batch Analyst* finds a path in the graph, the authors iterate through the edges to distinguish between different modes of travel. For each edge, the application firstly identifies the origin node and according to its attributes, it is possible to distinguish between private vehicle legs and public transport legs. The process then distinguishes between the modes of travel to calculate cost/fares for these modes differently.

5.3.1 Private vehicles

According to the Automobile Association [24] the Total Operating Cost of private vehicles consists of Running Cost and Fixed Cost components. The Fixed Cost component includes vehicle finance (cost of capital), insurance, vehicle tracking and licenses. In the AA model the Fixed Cost is estimated from the purchase price and annual distance travelled. The Running Cost component includes fuel, maintenance (service and repair) and tyre wear. In

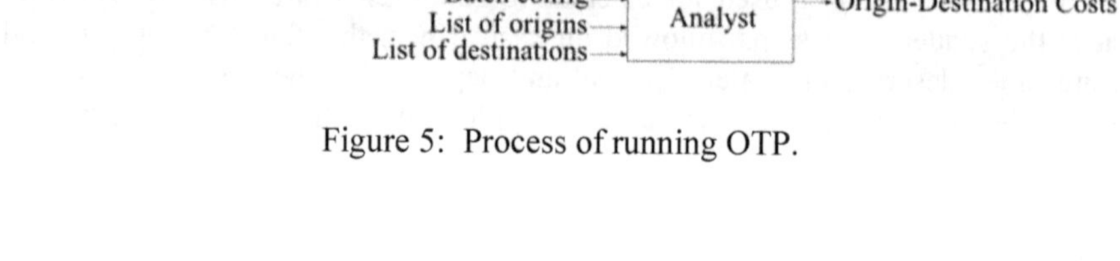

Figure 5: Process of running OTP.

the AA model the Running Cost is estimated based on the engine capacity and current fuel price. Assuming an average vehicle purchase price of R225,000-00 (incl. VAT), average distance travelled of 27,500 km per annum, average engine capacity of 1600 cm^3 and representative fuel price of R13.50 per litre, the Total Operating Cost per km comes to:

Fixed Cost	R 2.40	62%
Running Cost	R 1.48	38%
Total Operating Cost	R 3.88	100%

There is a significant body of empirical knowledge on how fuel consumption varies with network and traffic conditions. In the above example the fuel cost alone represents 28% of the Total Operating Cost. The empirical results point to a simple relationship between the fuel consumption F (litre/100 km) and the average speed of the entire trip V in km per hour (total distance D divided by total trip duration T).

$$F = k_1 + \frac{k_2}{V}, \tag{2}$$

recognising that $V = D/T$, this can easily be converted to a fuel cost C_F (Rand per km) using the trip distance D (km) and fuel price P_F (Rand per litre) as follows:

$$C_F = P_F F, \tag{3}$$

$$= P_F \left(k_1 + \frac{Tk_2}{D}\right). \tag{4}$$

Due to this relationship, the fastest route found by OTP will most likely also be the lowest-cost route.

5.3.2 Transit

For a transit-only configuration, all nodes are disregarded in the path until a node ID is found that contains the description "depart". This indicates the start of the transit trip. All subsequent edges are considered until a node is found that contains the description "arrive", which indicates the end of the transit trip. Once both the departure and arrival nodes have been found, the cost for the leg is obtained from the fares table in the agency's GTFS feed. If more transit legs are found in the path, for example due to a transfer, the fares are added to determine the total fare for the itinerary.

6 DATA PREPARATION

The work described in this section was taken from the validation phase of the two urban simulation projects as case studies for the research and includes the Cities of Tshwane and Nelson Mandela Bay. The objective during this phase is to create a representative transit network for the period between 2001 and 2011 which was used for validating UrbanSim and OTP as described in Subsection 7.2 hereunder. With that as objective, it was possible to use much simpler versions of the GTFS feed compared to what would be published to GTFS Exchange. Of course it would have been an advantage to start with real GTFS feeds but in South Africa only a few public transport operators use GTFS to publish their schedules. The authors consequently had no option but to set up simplified tables as described below for the City of Tshwane case study. Since only the morning peak of one weekday was used to simulate and to calculate all the measures of accessibility required by UrbanSim, many of the tables can be simplified to a fraction of the records that would normally be included.

Various transit operators exist in the City of Tshwane, which include bus services, minibus taxis and passenger rail services. In the research model the Gautrain High Speed Rail which only started operations in 2010, was not included.

6.1 Routes and stops

For some of the operators, it was possible to obtain route and stop information, which were matched, merged, and then allocated the various stops to routes. From this merged file, it was possible to extract all the necessary information to generate most of the GTFS files. In the case where no stop information was available, stops were generated on the route at 1 km intervals. The logic behind this was that for this purpose the research was to allow a walking distance of up to 2 km for people in the model to access public transport. Multiple stops within the 2-km radius would therefore make no difference to the resulting accessibilities. Although minibus taxis take the liberty of stopping almost anywhere they deem necessary, the same logic as above was applied and thus include options for stops at 1 km intervals to ease the computational burden.

6.2 Frequencies and travel time

To determine the times between stops, the research assumes different average travelling speeds for each of the modes and accordingly calculated the duration. For buses, a speed of 30 km/h was used, for minibus taxis a speed of 45 km/h, and for passenger rail a speed of 60 km/h. In the case where passenger rail, buses, and minibus taxis operate on a similar route between the same origin and destination, the passenger rail will be the most attractive since it will not only be the fastest but also the lowest cost due to the fare structure described below and in Subsection 4.3.2 above.

Since the research don't use actual schedules of the transit operators for modelling purposes, the service frequency was arbitrarily set to 60 seconds. This is clearly not representative of the real world since the research has shown that the majority of transit commuters are cost sensitive. The authors argue that in the real world commuters wait for the next service. This is to ensure that the OTP succeeds in finding a route on account of scheduling. The frequency has no influence, however, on the duration between stops; these are still dependent on the stop times that are calculated from the speed of the vehicles.

6.3 Fares

For passenger rail, a distance-based approach is used in which the trip cost increases with distance even though the rate per km decreases. For the research the rates given for different distance intervals and calculated a cost between each pair of stations are used. The fares ranged between R4.27 and R6.11 per trip. For bus services, a uniform trip-based fare of R8.75 per trip is used. For minibus taxis, a trip-based fare is applied. Where the methodology could not match fares with routes, a fare was allocated to the route based on the fare of a similar route. The fares ranged between R6 and R20 per trip.

6.4 Road network

The road network is generated from OpenStreetMap (OSM). In the research the Geofabrik GmbH [25] server is applied to obtain a snapshot of the OSM network for South Africa and Osmosis [26], a Java-based command-line application, to clip the network to the extent of the City of Tshwane. All default values of the network were kept, since the network is only used for connectivity purposes and not traffic volumes, etc.

6.5 Origin–destination matrix/zones

The research used the centroids of the zones described in Subsection 4.2.1 as origin and destination points and an R script to piece together separate output files produced by the *Batch Analyst* runs into one text file containing the origin, destination, OD distance on the road network and a number of monetary costs representing different mode choices aimed at specific UrbanSim sub models. These mostly include drive-alone, transit and a combined option.

7 RESULTS

7.1 Accessibility

In presenting the results of accessibilities calculated by UrbanSim from the OD matrix provided by OTP, the authors have chosen to present a measure that is much more intuitive than the logsum (eqn (1)) used by UrbanSim. Such a measure is the number of jobs that can be reached by limiting expenditure on transportation to the maximum of household affordability. This limit has rather arbitrarily been taken as 30% of household income, simply because the literature in South Africa abounds with concerns about low-income households having to spend that much on transportation.

The following two maps (Figs. 6 and 7) show the number of formal non-home-based jobs that could be reached in 2001 from any origin in the City of Tshwane without spending more than 30% of household income on transportation by different modes of transport. The origins (and destinations used by OTP) represent the zones explained in Subsection 4.2.1 above. The location of jobs was taken from the base year of the validation phase to be discussed in the next section. The first map (Fig. 6) shows the number of jobs that can be reached by private vehicle in drive-alone mode and the second map (Fig. 7) shows the number of jobs that can be reached by rail, bus or minibus taxi.

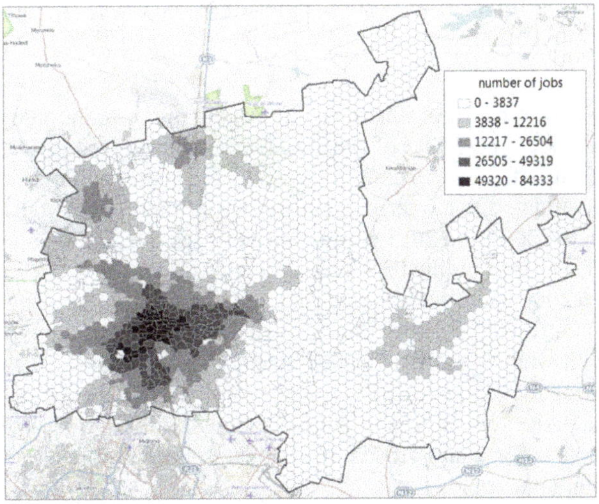

Figure 6: Number of jobs that could be reached in 2001 by spending less than 30% of household income on transportation by private vehicle.

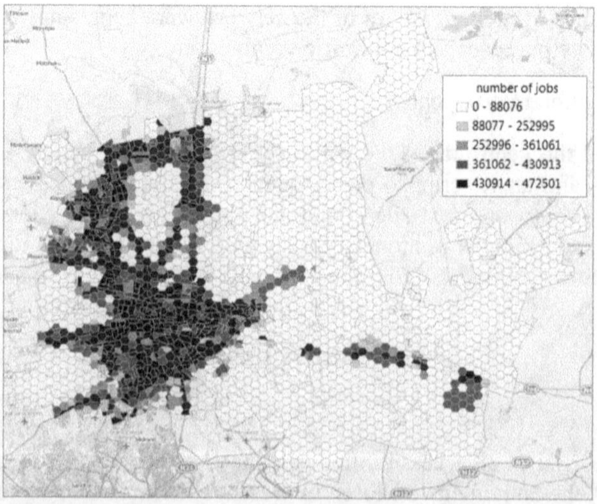

Figure 7: Number of jobs that could be reached in 2001 by spending less than 30% of household income on public transportation.

Although the maps show the enabling economic role of public transportation by providing access to jobs in many more locations, the real impact is appreciated when the number of jobs is considered that can be accessed as a percentage of the total number of jobs available in the city. By *'available'*, the research does not imply that the job is vacant or that a job seeker has the necessary qualifications, it simply means that if all other criteria are met, excessive transportation costs will not prevent someone from accessing an opportunity. If it is assumed that the average number of jobs that can be accessed from the 100 best placed origin zones (out of 2690) as a percentage of the total number of non-home-based jobs (664,395) in 2001 the result varies from 8% with no public transport to 71% with the representative public transport used for model validation.

7.2 Model validation

In this section the research output illustrates the results of the preliminary validation of the Urban Simulation Model in the City of Nelson Mandela Bay case study, which is a bit closer to finality than the City of Tshwane case study model at the time of reporting on the research.

All models are validated as a matter of course by simulating a period in the past, such as the period between the 2001 and 2011 population censuses released by Statistics SA. For this purpose the model is trained on the base year data and subsequently the predictions from the simulation between 2001 and 2011 are used for validation. The authors then compare various measures of urban growth, such as the growth in number of households forecast by UrbanSim/OTP with what actually happened during the same period according to Statistics SA at a resolution of the zones described in section 4.2.1 above. Because this involves a correction to the statistics for changes in sub-place boundaries between censuses, the authors also applies a normalisation against a commercial dataset derived by Geo Terra Image (GTI) from high resolution imagery.

The results are shown in Fig. 8 as a comparison of the growth in households forecast by UrbanSim/OTP to the actual growth in households according to Statistics SA (left) as well as a comparison of the actual growths according to Statistics and GTI (right). Both graphs exhibit a number of serious outliers. Upon investigation of the top 20 outliers produced by UrbanSim, are all found to have logical explanations. For example a township is established in the year 2000. UrbanSim sees undeveloped prime property with a great potential return on investment and develops the area to its full potential in the next 10 years. Meanwhile in the real world there is a pending land-restitution claim on the land which prevents any development from taking place. There are various ways of informing UrbanSim about such properties, provided of course that it is known.

If the top 20 outliers are corrected, a comparison of the remaining points in Fig. 8 show that the predictive accuracy of the model is comparable to a comparison of the actual growth measured by different methods employed by two reputable organisations. The predictive accuracy of the model is estimated at about 2 housing units per hectare. While it is conceded that there is a 3-year difference in duration, the authors regard the results as highly satisfactory as an overall validation of UrbanSim and OTP used together as described in this chapter.

8 DISCUSSION

In the motivation for adopting measures of accessibility based on the monetary cost of commuting, which does not depend on the congested state of the network, the research departure focuses on the potential for substantial reductions in overall simulation time because it essentially replaces 5 to 30 runs of MATSim (depending on extent of austerity measures to limit interaction with UrbanSim) with a few runs of OTP (one for the base year and then only when substantial changes are introduced to the network). The reduction in simulation time for 5 to 30 runs of MATSim is not being measured in hours but in one or more days. Since OTP requires no input from UrbanSim, all the OTP work can be completed before UrbanSim starts, even on a different computer. UrbanSim has great functionality for managing scenarios and can easily be configured to calculate the accessibilities from different OTP OD matrices representing the simulation year in which significant network changes are to be introduced.

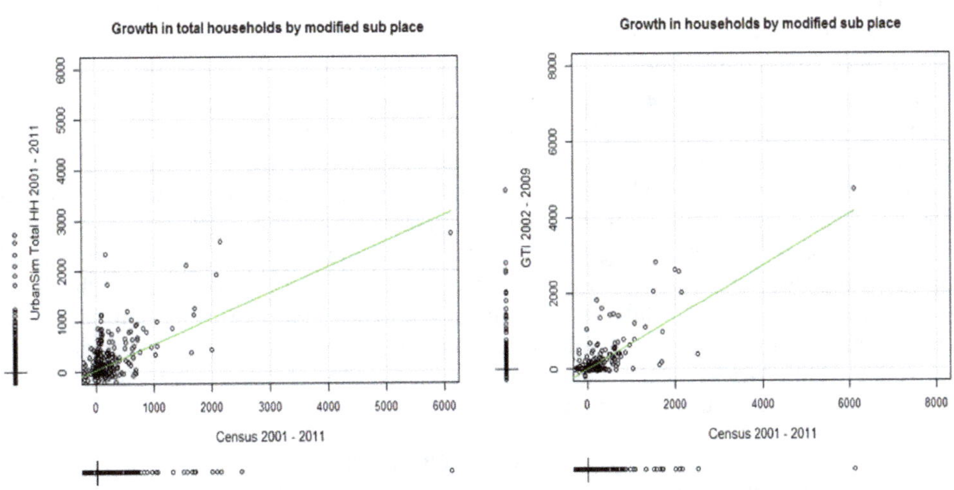

Figure 8: Model validation by comparison with actual growth according to Statistics SA (left) and actual growth according to GTI (right).

Apart from the reduction in computing time the outcome is also superior by virtue of the fact that UrbanSim is able to calculate the accessibility measures based on the correct location of jobs for every year of the simulation. Consider for example two UrbanSim models in which one interacts with MATSim as depicted in Fig. 1 (above) and the other with OTP. In both cases the research is restricted in interaction to once every 10 years. In the first case MATSim will only be able to calculate accessibilities based on the correct location of households and jobs every 10 years and UrbanSim will have to use these for the next 9 years even though the locations are not static. This is not a limitation in the second case because OTP provides the travel cost per OD pair and UrbanSim is able to calculate accessibilities based on the correct location of jobs, also in the 9 years between interactions.

While the research has focussed on the advantages of using OTP the authors would also like to point out some disadvantages. The relative worth of the different scenarios that may be simulated for a city is usually assessed by comparing a number of indicators. If these indicators include variables such as travel time or carbon emissions from vehicles, OTP will not provide the answers because it does not model traffic volumes and congestion. In such cases the practitioner would still have to run a transportation model, but this could be restricted to one run at the start and end years of the simulation.

9 CONCLUSION

This chapter describes the complex nature of urban systems and examines the role of Land Use Transportation Interaction (LUTI) models in supporting the equally complex urban planning processes that shape our cities. The research supports the LUTI model that is being used to simulate spatial growth patterns in South Africa to improve understanding of the future demand for infrastructure, facilities and services inclusive of water, sanitation, electricity and transportation. This model is based on adapted versions of UrbanSim and OpenTripPlanner (OTP).

The authors present a case for using the monetary cost of commuting as an acceptable proxy for generalised cost because the majority of households in South Africa, and possibly in other developing countries, tend to be more sensitive to the cost of a trip than the duration. Since the monetary cost of commuting does not depend on the congested state of the transportation network, the research output illustrated how to modify OTP to calculate lowest-cost commuting trips by various modes of transport and provide the costs to UrbanSim for calculating measures of accessibility to jobs, which could in turn influence household location choices. In the process the research achieves substantial improvements in data preparation and simulation times. It validates the combined UrbanSim and OTP model by simulating a ten year period in the past, compares the forecasts with the actual growth in households according to two reputable sources and obtains very favourable results as output.

The research further highlights the benefits of using the General Transit Feed Specification and encourage more transit operators to follow the example of those that have already published their schedules as a means of marketing and improving their service to commuters, promoting tourism by allowing international visitors to plan itineraries before arriving in South Africa and allow more applications to be developed to inform and improve the daily experience of transit commuters.

ACKNOWLEDGEMENT

The authors gratefully acknowledge the financial support of the Department of Science and Technology (stepsa.org).

REFERENCES

[1] Cervero, R., Linking urban transport and land use in developing countries. *The Journal of Transport and Land Use*, **6**(1), pp. 7–24, 2013.

[2] United Nations, The millennium development goals report 2011. Technical report, United Nations, New York, 2011, http://www.un.org/millenniumgoals/reports.shtml

[3] Statistics South Africa, Census 2011, http://www.statssa.gov.za/census2011/default.asp

[4] Wegener, M., Overview of land-use transport models. *Transport Geography and Spatial Systems. Handbook 5 of the Handbook in Transport*, eds D.A. Hensher and K. Button, Pergamon/Elsevier Science: Kidlington, UK, pp. 127–146, 2004.

[5] Wegener, M., Simulation of land use and transport: State of the art. *SustainCity Conference*, ETH Zurich, 17–18 April 2013.

[6] Waddell, P., Confronting the bane of endogeneity in modelling urban social dynamics. *Workshop on Modelling Urban Social Dynamics*, University of Surrey, 2005.

[7] Wegener, M., Operational urban models: State of the art. *Journal of the American Planning Association*, **60**(1), pp. 17–29, 1994.

[8] Simmonds, D., Waddell, P. & Wegener, M., Equilibrium v. dynamics in urban modelling. *Paper presented at the Symposium on Applied Urban Modelling (AUM 2011) "Innovation in Urban Modelling" at the University of Cambridge*, 23–24 May 2011.

[9] Waddell, P. & Ulfarsson, G.F., Introduction to urban simulation: Design and development of operational models. *Handbook in Transport, Volume 5: Transport Geography and Spatial Systems*, eds Stopher, Button, Kingsley & Hensher, Pergamon Press, pp. 203–236, 2004.

[10] McFadden, D., Conditional logit analysis of qualitative choice behavior. *Frontiers in Econometrics*, ed. P. Zarembka, Academic Press: New York, pp. 105–142, 1974.

[11] McFadden, D., Econometric models of probabilistic choice. *StrucAppendix A: Data Preparation in a Sparse Data Environment*, eds C. Manski & D. McFadden, 1981.

[12] Nicolai, T.W. & Nagel, K., Investigating accessibility indicators for feedback from a travel to a land use model. *SustainCity Working Paper*, 6.2, TU Berlin, 2011.

[13] Awaludin, A. & Chen, D., UrbanSim Parallel Programming Capstone Paper University of Washington, *Computer Science and Engineering 481E*, June 4, 2007.

[14] Raney, B. & Nagel, K., Truly agent-based strategy selection for transportation simulations, 2002, http://www.ltrc.lsu.edu/TRB_82/TRB2003-002258.pdf

[15] Zöllig Renner, C., Schirmer, C.P. & Müller, K., Case Study Zurich Simulation, Zurich, April 2013.

[16] Hansen, W., How accessibility shapes land use. *Journal of the American Planning Association*, **25**(2), pp. 73–76, 1959.

[17] Moeckel, R., *Business location decisions and urban sprawl – a microsimulation of business relocation and firmography*, PhD thesis, Department of Spatial Planning at the University of Dortmund, 2006.

[18] Visagie, J., Who are the middle class in South Africa? Does it matter for policy? 2013, www.econ3x3.org/article/who-are-middle-class-south-africa-does-it-matter-policy#sthash.lXzwvRCb.dpuf

[19] Visagie, J. & Posel, D., A reconsideration of what and who is middle class in South Africa. *ESSA Conference*, University of Kwazulu-Natal, 2011.

[20] National Household Travel Survey (NHTS), Statistical Release PO320, 2013.

[21] Google Developers, The General Transit Feed Specification Reference, https://developers.google.com/transit/gtfs/reference

[22] TriMet 2015, Map Trip Planner, http://trimet.org/howtoride/maptripplanner.htm
[23] GTFS Data Exchange, http://www.gtfs-data-exchange.com/
[24] Automobile Association, http://www.aa.co.za/on-the-road/calculator-tools/rates-calculator.html
[25] Geofabrik GmbH, http://download.geofabrik.de/africa/south-africa-and-lesotho.html
[26] Osmosis 2015, http://wiki.openstreetmap.org/wiki/Osmosis

THE USE AND APPLICATION OF VECTOR OPTIMISATION METHODS FOR LAND USE AND TRANSPORTATION INTEGRATION

ILSE MARIA SCHOEMAN
North-West University (Potchefstroom), South Africa

ABSTRACT

The aim of this chapter is to use vector optimisation methods to create a user-friendly scorecard to measure on a local municipality level the outcomes of integration between land use and transport. The development and application of a scorecard inspires continuous assessment and improvement of services to determine the weak and strong points of each local municipality. Thereby, supporting decision making to intervene earlier at local municipalities which perform poorly. This will be done by a detailed discussion of an "improved" Saaty's method and an "improved" row geometric mean method to derive the relevant weights of the criteria and factors. Furthermore, by using two analytical processes (i.e. weighted sum process and the exponential weighted process) the ranking of the different local municipalities' performance can then be determined. From this application the level of integration between land use and transportation systems can be scientifically calculated.

Keywords: weights, scorecard, vector optimisation, multi-objective optimisation, multi-criteria analysis.

1 INTRODUCTION

Vector optimisation i.e. multi-objective optimisation, is the process used to optimise a set of objective functions of Marler et al. [1]. These concepts develop from the following areas:

- pure mathematics
- economic (welfare and equilibrium) theories
- game theory.

The solution of a vector optimisation problem is called a Pareto optimal set (i.e. a set of points). The multi criteria analysis (MCA) process, which is a special case of the general vector optimisation process, is based on a hierarchical aggregation procedure combined with an analysis function (like the linear additive method).

The analytic hierarchy process (AHP) see Saaty (1980) [2], Saaty (1990) [3], Saaty (2001) [4], Saaty et al. [5] and Vargas [6], is the best known MCA process, allowing the user to measure the consistency of the preference (of relative importance of the different criteria or factors) in a formal systematic way that will reduce the complexity and subjectiveness. In this chapter consistent pairwise comparison matrix refer to a transitive, positive, square, reciprocal matrix, Saaty (1980) [2].

The AHP performs pairwise comparisons

- to calculate the relative importance of criteria;
- transforms subjective judgments into objective measures;
- to deal with complex situations by using a realistic model which "include" and "measure" all important tangible and intangible quantitatively and qualitative factors.

The reason for this study arise from South Africa's main legislative and regularity framework that dictate and define the Municipalities Internal Auditing processes and which also form part of the bigger performance management system. The relevant legislation and regulations can be summarised as follows:

- **Municipal Systems Act, 2000, Chapter 6, Section 45** requires that the *results of performance measurements, be audited* as part of the Municipality's Internal Auditing processes.

- **The Municipal Planning and Performance Management Regulations, regulation 14 (1), 2001**, of the Systems Act, states that it is the Municipality's responsibility to *develop and implement mechanisms, systems and processes* for the auditing of the performance measurements, as part of its internal audit processes.

- **The Municipal Finance and Management Act No 56 of 2003, section 165 (2) (b) (v)** requires that the Internal Audit Unit of a Municipality, *report to the Audit Committee on matters relating to Performance Management.*

- **Performance Information Public Audit Act, Notice 464, 2007**, gives the framework for managing programme performance information.

The performance committee responsibilities with respect to municipalities involve the following:

- **review** the quarterly reports submitted by the municipality's internal audit units;
- **review** the municipality's performance management system;
- **assess** whether the performance indicators are sufficient;
- make **recommendations** on the municipality's performance management system.

The following problems were encountered with the current performance management system:

1. Factors or indicators were not measurable;
2. Lack of formal processes or systems to monitor the achievement of objectives on municipality level;
3. Lack of formal processes with actions on how to address respective factors with low scores. Thus, there is a lack of specific actions to improve poor performance on a specific factor;
4. Lack of formal penalty systems with repercussions for individuals because of performance reports that are late or never submitted or because of poor performance;
5. Traceable data or source documentation, hence a transparent measuring process with accurate, reliable and valid factor scores;
6. Lack of substantial reasons for non-performance.

To solve these problems, criteria and factors (i.e. Key Performance Indicators) must not be determined or manipulated by municipalities. A scientific process must be followed in the determination of criteria and factors and there must be different factors for different municipalities. Therefore a municipality class may depend on the size and function of the city or town. The scoring process is currently too simplistic (only three outcome levels) with too many factors to be measured and the analysis process is insufficient. Currently the analysis process gives an overall score without pointing out the strong and weak points of each municipality. To sidestep manipulation of data and information, monthly data and source documentation must be given to an external agency which then must calculate, monitor and measure the performance of each municipality. Hence, each municipality must only know the criteria on which they will be measured but the respective factor definitions that will be used and how each factor is calculated must be unknown to the municipalities.

In this chapter the author will create a user-friendly scorecard that can be used by local municipalities (LM) to optimise the application of land use and transportation integration goals and objectives on a local municipal (LM) level. There is a need for land use and transportation integration due to the

- fragmented urban form;
- long commuting times;
- high travelling cost.

2 METHODOLOGY

Koski et al. [7] reduces the number of original objective functions by grouping the objective functions into sets i.e. criteria with common characteristics; by doing so the main goal (proxy by objectives) is broken down into criteria which are evaluated in terms of their respective factors. The different criteria can be measured and its application and functionality assessed in terms of its respective factors.

The methodology applied is to use a vector optimisation method to determine how to optimise the different criteria which are linked to the milestones of land use and transportation integration. Note that this chapter is an extension of the work done by Schoeman in [8] where the author only considered steps 1 to 4b) in the Methodology given in Fig. 1.

2.1 The applicable criteria and factors

Manual [9], United Nations [10] and Niven [11] provide the following guidance applicable to the assessment characteristics for any set of criteria, factors or Key Performance Indicators (KPI):

- **broad in coverage of all aspects** of land use and transportation integration but also **mutually disjoint** (i.e. no double counting) and of good quality; For more information on mutually disjointness refer to Appendix A.
- **relationship** to the milestone;
- **data** on these criteria or factors must be **available** and **easily obtainable**;
- **usable or relevant** on local municipality level for land use and transportation integration;
- **measurable** in a clear and understandable qualitative or quantitative way.

Schoeman [8] identified the applicable objectives as densification, mixed land use activities, the enforcement of land use and traffic policies, accessibility, mobility and centralised data on transportation and land use including the applicable GIS files. These objectives satisfy the above properties for the realisation of land use and transport integration on local municipality level.

Next, follows the definitions of the respective factors for the alternative criteria. Note that some of these definitions differ from the ones defined by Schoeman [8] which is due to adjustments made during the scoring stage when problems (like mutual dependence or inaccessible data) were discovered.

Let the binary value (BV) function of the variable X_i be given as

$$BV(X_i) = \begin{cases} 1 & \text{if} \quad X_i = \text{Yes}; \\ 0 & \text{if} \quad X_i = \text{No}. \end{cases}$$

(1) *For the criterion* ***density*** the following factors were considered

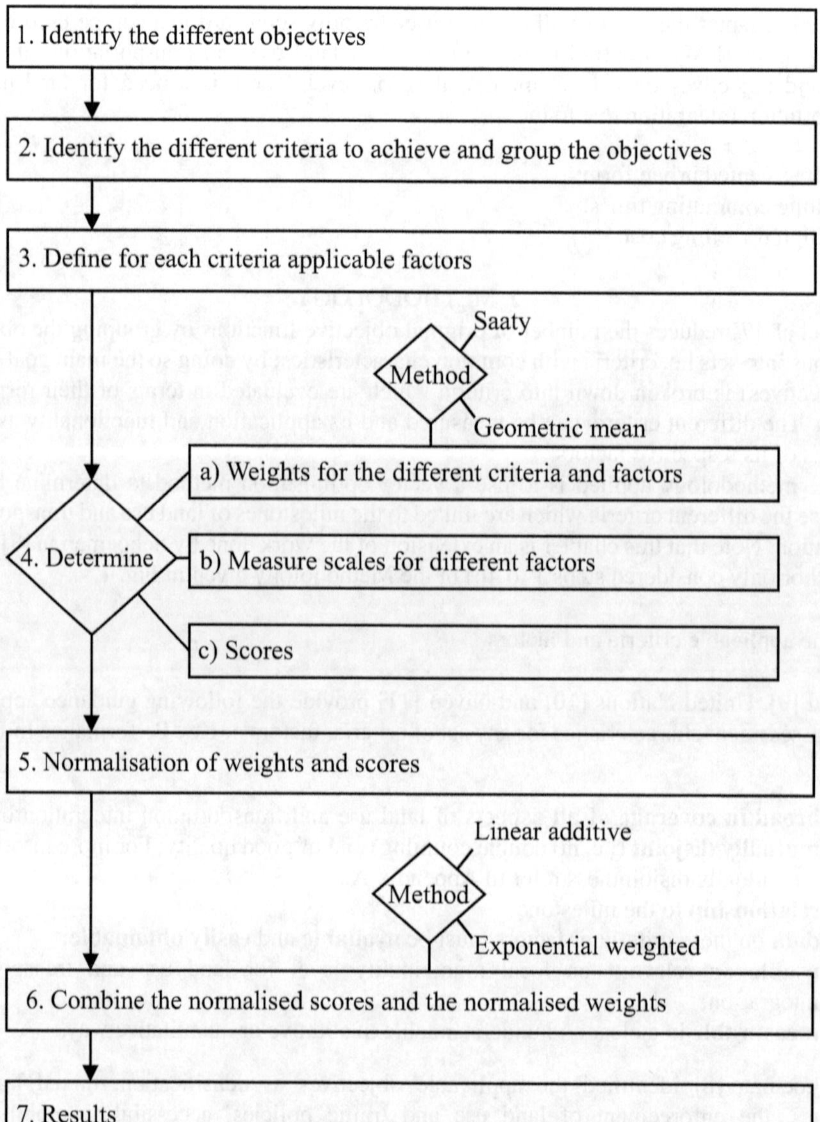

Figure 1: Methodology for the vector optimisation method. (*Source: own construction.*)

(1.1) *Gini concentration ratio* (GCR) of the neighbourhoods in the local municipality. This measure is adapted from the work of Gini [12] where it was mainly used in economics for the comparison of the income distribution of different nations. In this chapter the Gini concentration ratio is used to describe the distribution of the urban population in the different residential neighbourhoods in relation to the area. Hence, this ratio will be the instrument which will contribute to the development of sustainable and resilient urban neighbourhoods by constraining too much sprawling and also constraining neighbourhoods that are too compact in comparison with the other. The GCR is calculated as follows

$$GCR = \frac{\sum_{i=1}^{n}(Y_{i+1}X_i) - \sum_{i=1}^{n}(X_{i+1}Y_i)}{10\,000}$$

Y_i = Cumulative **area** proportion of each residential neighbourhood;
X_i = Cumulative **population** proportion of each residential neighbourhood;
n = The # of residential neighbourhoods in the local municipality.

With the factor classification defined as

$$S_{Gini} = \begin{cases} \text{- Maximal inequality among} & GCR = 1; \\ \text{the neighbourhoods densities;} & \\ \text{Penalty function} = 1 - GCR & \\ \text{- Complete equality among} & \\ \text{the neighbourhoods densities} & GCR = 0. \\ \text{Penalty function} = 1 - GCR & \end{cases}$$

From this classification it follows that if the GCR converges to zero, the better the density criteria of a city, therefore the penalty function was used to transform these values to a scale corresponding with consistent meaning among the scores of the different factors where scores that converge to one implies a positive effect on land use and transport integration and scores that converge to zero implies a less positive or no contribution to land use and transport integration.

(1.2) The *Employment-population density* of the local municipality. The proxy for this factor is the Employment rate defined as the proportion of the working-age population that is employed (see StatsSA [13]).

(2) *The criterion **enforcement of land use and transport*** (LUT) policies

(2.1) *Integration of the three spheres of government* (IG) factor.
Let IDP be the Integrated Development Plan of the applicable municipality and define $S_{integration} = \sum_{i=1}^{5} BV(X_i)$ where

X_1 = Linkages in the **planning** between the 3 spheres of government and also the IDP;
X_2 = Linkages in the **projects** between the 3 spheres of government and also the IDP;
X_3 = Linkages in the **budgets** between the 3 spheres of government and also the IDP;
X_4 = Integrated monitoring of **expenditure** on all three spheres of government;
X_5 = Integrated monitoring if programs/projects are finished **on time and within projected budget.**

Hence, the IG factor was classified in terms of the value of $S_{integration}$ i.e.

$$IG_{factor} = \begin{cases} 1 & \text{if } S_{integration} = 5; \\ 0.75 & \text{if } S_{integration} = 4; \\ 0.5 & \text{if } S_{integration} = 3; \\ 0.25 & \text{if } S_{integration} = 2; \\ 0.1 & \text{if } S_{integration} = 1. \end{cases}$$

(2.2) *Database factor* (S_D):
Let

$$S_{database} = \sum_{i=1}^{5} BV(Z_i)$$

for the evaluation of the database on LM level. Consider if the LM database has the following properties

Z_1 = Consistently updated and accessible to the public;
Z_2 = Integrated database between the various spheres of government;
Z_3 = Contain GIS data on transport network and time series data on **transport demand**;
Z_4 = Contain GIS and time series data on **land use**;
Z_5 = Contain GIS and time series data on **engineering services.**

Furthermore, the range of the rating scale for the Database factor is given the value

$$S_D = \begin{cases} 1 & \text{if } S_{database} = 5; \\ 0.75 & \text{if } S_{database} = 4; \\ 0.5 & \text{if } S_{database} = 3; \\ 0.25 & \text{if } S_{database} = 2; \\ 0.1 & \text{if } S_{database} = 1. \end{cases}$$

(3) *For the criterion **accessibility***. The authors in Nicolai et al. [14] consider the following accessibility variables for developed countries (i.ethe accessibility variables used in UrbanSim & MATSim):

(1) Travel cost (monetary value);
(2) Travel distance;
(3) How affordable the rent or bond payments are;
(4) Congestion levels;
(5) Travel time.

In Chapter 3 of this book the authors argued that in developing countries where a large proportion of the households are poor, travelling cost is the most important accessibility variable and travel time the least important. Also, the affordability of rent or bond payment were not considered as an accessibility variable since this is not applicable on developing low-income study area where most of the houses are RDP-housing or informal housing.
Hence, the factors for developing countries considered under the accessibility criteria are as follows

(3.1) *Travel Cost* (TC) which is measured as average daily travelling cost (% of income) for the applicable city or town; Let PC be the average provincial travel cost % and NC be the average national travel cost %. Then

$$S_{TC} = \begin{cases} 1 & \text{if } TC \overset{5\%\ or\ more}{<} (PC \text{ and } NC); \\ 0.75 & \text{if } TC \overset{less\ than\ 5\%}{<} (PC \text{ and } NC); \\ 0.5 & \text{if } TC \overset{5\%\ or\ more}{<} (PC \text{ or } NC); \\ 0.25 & \text{if } TC \overset{less\ than\ 5\%}{<} (PC \text{ or } NC); \\ 0.1 & \text{if } TC > (PC \text{ and } NC). \end{cases}$$

Therefore, if the travel cost for Tlokwe $= 18\%$, the North West Province travel cost $= 15\%$ and the average National Travel cost is equal to 20%. It follows from this information that

$$PC \overset{3\%}{<} TC \overset{2\%}{<} NC.$$

Hence, $S_{TC} = 0.25$. Note that ideally the monthly household transport expenditure should be less than 10% of income, but with the majority of the working population in South Africa classified as low income earners this percentage is only a dream.

(3.2) *Public transport (referring to minibus taxi, train or bus) usage* =

$$\frac{Public\, transport\; usage\; in\; LM\; area}{Population\; in\; LM\; area}$$

(3.3) *Average distance between residential neighbourhoods and area of employment* $:= d(Res, work)$. Consider the average proximity to work from various informal residential neighbourhoods.

$$d(Res, empl) = \begin{cases} 1 & \text{if } d(Res, work) < 2.5\,\text{km}; \\ 0.75 & \text{if } d(Res, work) \in [2.5\,\text{km}; 5\,\text{km}); \\ 0.5 & \text{if } d(Res, work) \in [5\,\text{km}; 7.5\,\text{km}); \\ 0.25 & \text{if } d(Res, work) \in [7.5\,\text{km}; 10\,\text{km}]; \\ 0.1 & \text{if } d(Res, work) > 10\,\text{km}. \end{cases}$$

These scale intervals for the different factors originate from the Guideline of the Department of Public Service and Administration see DPSA [15] but are much more refined;

(3.4) *Travel Times* (TT) = Average total travel time (i.e. walking, waiting and time travelling) per day for/on public transport; Let PV be the average provincial travel time value for the applicable public transport mode and NV the applicable national value. Then

$$S_{TT} = \begin{cases} 1 & \text{if } TT \overset{30\ min.\ or\ more}{<} (PV \text{ and } NV); \\ 0.75 & \text{if } TT \overset{between\ 10\ min.\ and\ 30\ min.}{<} (PV \text{ and } NV); \\ 0.5 & \text{if } TT \overset{less\ than\ 10\ min.}{<} (PV \text{ and } NV); \\ 0.25 & \text{if } TT > (PV \text{ or } NV); \\ 0.1 & \text{if } TT > (PV \text{ and } NV). \end{cases}$$

(4) *For the criterion* ***mobility*** consider

(4.1) *Average traffic congestion levels*
The scale intervals for the average (peak hour) traffic congestion i.e. volumes to capacity (i.e. V/C) ratio, follows from the guidelines given in any Traffic Engineering Manual.

$$\text{Congestion level} = \begin{cases} 1 & \text{if } Avg(V/C) < 0.9; \\ 0.75 & \text{if } Avg(V/C) \in [0.9; 0.94); \\ 0.5 & \text{if } Avg(V/C) \in [0.94; 1.00); \\ 0.25 & \text{if } Avg(V/C) \in [1.00; 1.5); \\ 0.1 & \text{if } Avg(V/C) \geq 1.5. \end{cases}$$

(4.2) *Forecasted (15-year) traffic congestion value*
The 15-year forecasted average peak hour V/C ratio (i.e. $\mathbf{E}(V/C)$). To calculate the forecasted congestion levels, use the population growth rates and data on the modal selection.

$$\text{Forecasted congestion levels} = \begin{cases} 1 & \text{if } \mathbf{E}(V/C) < 0.9; \\ 0.75 & \text{if } \mathbf{E}(V/C) \in [0.9; 0.94); \\ 0.5 & \text{if } \mathbf{E}(V/C) \in [0.94; 1.00); \\ 0.25 & \text{if } \mathbf{E}(V/C) \in [1.00; 1.5); \\ 0.1 & \text{if } \mathbf{E}(V/C) \geq 1.5. \end{cases}$$

(4.3) *Road safety* (RS). This is measured by considering the number of fatalities per 1000 inhabitants in LM area per annum. Note that the South African per annum road fatalities $= \frac{0.22\ fatalities}{1000\ inhabitants}$ and Australia's per annum road fatalities $< \frac{0.036\ fatalities}{1000\ inhabitants}$ (see Iaych et al. [16]).

$$RS = \begin{cases} 1 & \text{if } \textit{fatalities per } 1000 \textit{ inhabitants} < \frac{0.06}{1000}; \\ 0.75 & \text{if } \textit{fatalities per } 1000 \textit{ inhabitants} \in [\frac{0.06}{1000}; \frac{0.09}{1000}); \\ 0.5 & \text{if } \textit{fatalities per } 1000 \textit{ inhabitants} \in [\frac{0.09}{1000}; \frac{0.12}{1000}); \\ 0.25 & \text{if } \textit{fatalities per } 1000 \textit{ inhabitants} \in [\frac{0.12}{1000}; \frac{0.15}{1000}]; \\ 0.1 & \text{if } \textit{fatalities per } 1000 \textit{ inhabitants} > \frac{0.15}{1000}. \end{cases}$$

The different ratios in the intervals coincide with the ratios in the classes used to compare the road fatalities in the world.

2.2 The weights of the respective criteria and factors

The different (criteria and factors) weights were determined from literature and by using the prioritisation procedures of Saaty and the Row Geometric Mean Method (RGMM).

A summary of Saaty's method, constructed from the work of De Jong [17], Saaty (1977) [18], Saaty (1980) [2], Saaty (1984) [19], Saaty (1990b) [20], Saaty (2005) [21] and Tomashevskii (2015) [22] (see Appendix B for the iterative algorithm):

(1) Saaty's scale for the starting preferences (i.e. the starting values, $a_{ij}, \forall\ i, j \in [1; m]$ of matrix A) are given in Table 1. These values describe the priority (order) of the (current and future) contribution (costs, benefits and risk) of the relevant criteria (or factors) to the objectives, which are determined using applicable literature (i.e. articles, policies and legislation) and if necessary surveys (public participation input) and interviews with experts.

From the starting priorities the reciprocal, square, positive, $(m \times m)$ Analytic Hierarchy Process (AHP) matrix,

$$A = \{a_{ij}\} \text{ with } i, j \in [1; m]$$

is calculated. Note that A is calculated using $m[m-1]/2$ pairwise comparisons in determining the importance of the different criteria or factors.

(2) Determine λ_{max} the largest eigenvalue of A.

Table 1: Saaty's scale of the starting preferences.

Starting priority value	Description of score
$\frac{1}{9}$	Extremely lower contribution
$\frac{1}{8}$	Very, very lower contribution
$\frac{1}{7}$	Demonstrated lower contribution
$\frac{1}{6}$	Strongly plus lower contribution
$\frac{1}{5}$	Strongly lower contribution
$\frac{1}{4}$	Moderately plus lower contribution
$\frac{1}{3}$	Moderately lower contribution
$\frac{1}{2}$	Weakly lower contribution
1	Equal contribution
2	Weakly higher contribution
3	Moderately higher contribution
4	Moderately plus higher contribution
5	Strong or essential contribution
6	Strong plus contribution
7	Demonstrated contribution
8	Very, very strong contribution
9	Extremely important contribution

(3) Calculate the values for the weights given by the eigenvector, $\mathbf{w}$ such that $A\mathbf{w} = \lambda_{max}\mathbf{w}$.
(4) Normalise the weights.
(5) To restrict the inconsistency and the subjectiveness of the judgements (or comparisons or preferences) Saaty derived the restriction of the contradictories in the form of the Consistency Ratio (CR_m) which is defined in terms of the Random Consistency Index (RCI_m) values which depend on the dimension, $(m \times m)$ of matrix A. By using a random generated sample of size 500 Saaty derived the following values

$$RCI_m \stackrel{[2]}{=} \begin{cases} 0.58 \text{ if } m = 3; \\ 0.90 \text{ if } m = 4; \\ 1.12 \text{ if } m = 5; \\ 1.24 \text{ if } m = 6; \\ 1.32 \text{ if } m = 7 \end{cases}$$

with

$$CR_m = \frac{\frac{\lambda_{max} - m}{m-1}}{RCI_m}.$$

Use Saaty's Consistency Ratio to test if A is consistent enough (c.e.)

$$A = \begin{cases} \text{c.e. if } CR_3 \leq 0.05 \text{ or } CR_4 \leq 0.08 \text{ or } CR_5 \leq 0.1 \text{ or } CR_6 \leq 0.2; \\ \text{inconsistent otherwise} \Rightarrow \text{revise A.} \end{cases}$$

Note from Saaty (1977) [18] that A is perfectly consistent if $\lambda_{max} = m$ and $CR_m = 0$ or if A is transitive i.e.

$$a_{ik} = \frac{a_{ir}}{a_{kr}}, \forall i, k, r \in [1, m].$$

A summary of Row Geometric Mean Method (RGMM) constructed using Aguarón et al. [23], Crawford et al. [24] and Tomashevskii (2014) [25]:

Note that the Row Geometric Mean Method is also known as the Logarithmic Least Squares Method.

(1) Determining the importance of the different criteria or factors by calculating A a reciprocally symmetric AHP matrix (see Appendix B).
(2) The normalized weights of geometric mean method is given by

$$w^* = [\overline{x_1}, \overline{x_2}, \ldots, \overline{x_m}]^t$$

with

$$\overline{x_k} = \frac{\left(\prod_{j=1}^{m} a_{kj}\right)^{\frac{1}{m}}}{\sum_{i=1}^{m}\left(\prod_{j=1}^{m} a_{ij}\right)^{\frac{1}{m}}}, \qquad k = 1, 2, \ldots, m.$$

(3) Test if A is consistent enough by considering the Row Geometric Consistency Index (RGCI) see Crawford et al. [24] given by

$$RGCI = \frac{2 \cdot \sum_{i<j}^{m} \log^2\left(a_{ik}\frac{\overline{x_k}}{\overline{x_i}}\right)}{(m-1)(m-2)}. \tag{1}$$

Here,

$$A = \begin{cases} \text{consistent enough if } RGCI \leq \mathit{Threshold}_{RGCI}; \\ \text{inconsistent otherwise} \Rightarrow \text{revise A.} \end{cases}$$

From Aguarón et al. [23] follows the following threshold values for the RGCI in the special case where $\overline{x_i} = \left(\prod_{j=1}^{m} a_{kj}\right)^{\frac{1}{m}}$, $k = 1, 2, \ldots, m$ in formula (1)

$$\text{Threshold}_{RGCI} = \begin{cases} 0.314 \text{ if } m = 3 \;\&\; CR = 10\%; \\ 0.352 \text{ if } m = 4 \;\&\; CR = 10\%; \\ 0.37 \text{ if } m > 4 \;\&\; CR = 10\%; \\ 0.1573 \text{ if } m = 3 \;\&\; CR = 5\%; \\ 0.1763 \text{ if } m = 4 \;\&\; CR = 5\%; \\ 0.185 \text{ if } m > 4 \;\&\; CR = 5\%; \\ 0.0314 \text{ if } m = 3 \;\&\; CR = 1\%; \\ 0.0352 \text{ if } m = 4 \;\&\; CR = 1\%; \\ 0.037 \text{ if } m > 4 \;\&\; CR = 1\%; \end{cases}$$

If A is inconsistent then it entails numerical errors, Δw_i of the weights, w_i.

Since the actual numerical errors are unknown the following problems exist with Saaty's method and the RGMM (see Tomashevskii (2015) [22], Tomashevskii (2014) [25], Bana et al. [26], Barzilai [27], Bozoki et al. [28], Johnson et al. [29], Karapetrovic et al. [30], Kwiesielewicz et al. [31] and Monsuur [32]):

- Rank reversal problem, i.e. the order of the weights calculated by Saaty method and RGMM differ;

- Right-left asymmetry i.e. $A\underline{\mu} = \lambda_{max}\underline{\mu} \quad \& \quad \underline{x}A = \lambda_{max}\underline{x} \quad \nRightarrow \underline{\mu} = \underline{x}$;
- Both methods are sensitive to outliers.

From Tomashevskii (2015) [22] and Tomashevskii (2014) [25] follows the following solutions to the above problems:

- Take the numerical errors also into account in the ordering of the weights therefore consider the ranking intervals, where

$$\text{ranking interval} := \text{normalised weight} \pm \text{numerical errors.}$$

- The ranking intervals must be mutually disjoint. Note that Tomashevskii (2014) [25], made a mistake in the calculation of the errors in his example. To see this consider the well-known hyperbolic identity $\cosh^2 x - \sinh^2 x = 1$ from this follows that $(w_i)^2 - (\Delta w_i)^2 = (w_i^*)^2\{(ch\Delta_i)^2 - (sh\Delta_i)^2\} = (w_i^*)^2$. Apply this identity to the values (which follows from eqns (14) and (15) in Tomashevskii (2014) [25]) in the example will lead to a contradiction.

Improved Method I: Combining Tomashevskii solutions and Saaty's method using Matlab (see Appendix C):

(1) Calculate the reciprocal square positive, $(m \times m)$ Analytic Hierarchy Process matrix,

$$A = \{a_{ij}\} \text{ with } i, j \in [1; m]$$

with starting values

$$a_{ij} \in \left\{\tfrac{1}{9}; \tfrac{1}{8}; \tfrac{1}{7}; \tfrac{1}{6}; \tfrac{1}{5}; \tfrac{1}{4}; \tfrac{1}{3}; \tfrac{1}{2}; 1; 2; 3; 4; 5; 6; 7; 8; 9\right\}.$$

Determine the starting priority (i.e. the order given by the usually inconsistent starting priority matrix) (see Appendix D).
Example: $criterion_m < criterion_2 < \cdots$.
Use the iterative algorithm (see Appendix B for the Matlab code) to improve (by using Saaty's idea) the consistency of the comparisons in the case of an inconsistent evaluation matrix, A. Calculate the end priority (i.e. the order given by the sufficiently consistent matrix). The decision maker must decide whether to adopt the sufficiently consistent matrix or not, by comparing the starting priorities and the end priorities.

(2) Compute the values for the weights given by the eigenvector, $\mathbf{w}$ such that $A\mathbf{w} = \lambda_{max}\mathbf{w}$, with λ_{max} the largest eigenvalue of A.

(3) Let $\overline{z} = [z_1, \ldots, z_m]^t$ be the corresponding normalised weights (eigenvector) associated with λ_{max}.

(4) Determine $\overline{\delta} := [\delta_1, \delta_2, \ldots, \delta_m]^t$ where

$$\delta_i \overset{[25]}{=} \sqrt{\frac{1}{(m-1)(m-2)} \sum_{k=1}^{m} \left(\frac{a_{ik} m z_k}{\lambda_{max}} - z_i\right)^2}, \qquad i = 1, \ldots, m$$

are the error terms of Saaty's method. See Appendix E for the evaluation of this formula.

(5) Consider the intervals $\mathbf{w}_S = \overline{z} \pm \overline{\delta}$. Mutually disjoint intervals, imply reliable weights.

Improved Method II: Combining Tomashevskii solutions and the Row Geometric Mean Method using Matlab:

(1) Calculate the consistent enough reciprocal square positive, $(m \times m)$ Analytic Hierarchy Process matrix, A (see Appendix B). Also, compute the starting and end priorities (as in Appendix D).

(2) Calculate the values of the normalised weights

$$\overline{x_k} = \frac{\left(\prod_{j=1}^{m} a_{kj}\right)^{\frac{1}{m}}}{\sum_{i=1}^{m}\left(\prod_{j=1}^{m} a_{ij}\right)^{\frac{1}{m}}}, \qquad k = 1, 2, \ldots, m.$$

(3) To eliminate all the problems it follows from the work of Tomashevskii [25] that it is better to consider $\mathbf{w} = [w_1, w_2, \ldots, w_m]^t$ with

$$w_i = \overline{x_i} \cdot \left(\frac{e^{\Delta_i} + e^{-\Delta_i}}{2}\right) = \overline{x_i} \cosh(\Delta_i)$$

where

$$\Delta_i = \sqrt{\frac{1}{(m-1)(m-2)} \sum_{k=1}^{m} \log^2\left(a_{ik}\frac{\overline{x_k}}{\overline{x_i}}\right)}.$$

Let $\varepsilon = [\varepsilon_1, \varepsilon_2, \ldots, \varepsilon_m]^t$ be the RGMM errors, with

$$\varepsilon_i = \overline{x_i} \cdot \left(\frac{e^{\Delta_i} - e^{-\Delta_i}}{2}\right) = \overline{x_i} \sinh(\Delta_i)$$

where

$$\Delta_i = \sqrt{\frac{1}{(m-1)(m-2)} \sum_{k=1}^{m} \log^2\left(a_{ik}\frac{\overline{x_k}}{\overline{x_i}}\right)}.$$

(4) Now, let the interval $\mathbf{w}_{RGMM} = \mathbf{w} \pm \varepsilon$.

$$Test_{Mutually\ disjointness} = \begin{cases} \text{Yes} \Rightarrow \textit{reliable weights}; \\ \text{No} \Rightarrow \textit{unreliable weights}. \end{cases}$$

For the calculation of the weights for the criteria the following Analytic Hierarchy Process (starting) matrix was used:

	Access	Mobility	Density	LUT
Access	1	1	2	3
Mobility	1	1	1.5	2.5
Density	0.5	0.67	1	1.5
LUT	0.34	0.4	0.67	1

From the Improved Method II follows the mutually disjoint intervals:

$$[0.3512; 0.3697], [0.3117; 0.3293], [0.1894; 0.1979] \text{ and } [0.1245; 0.127].$$

Hence, the criteria weights in Table 2 are reliable. The same path was followed in the determination of the factor weights.

Table 2: Weights.

Criteria and factor(s)	Norm. criteria weights	Norm. factor weights
1. Density	0.19	
1.1 Gini concentration ratio		0.1
1.2 Employment-population		0.09
2. Enforcement of LUT policies	0.13	
2.1 Integration		0.065
2.2 Database		0.065
3. Accessibility	0.36	
3.1 Travel Cost		0.211
3.2 Public transport usage		0.070
3.3 d(Res,Work)		0.059
3.4 Travel Time		0.020
4. Mobility	0.32	
4.1 Congestion		0.119
4.2 Forecasted congestion		0.118
4.3 Road safety		0.083

The main reason for the choice of these accessibility and mobility criteria, factor weights (AHP starting matrix) follows from Chapter 3 of this book where the authors argue that in a developing country (with large proportion of low income households) like South Africa (with 65% of low income households in Census 2011, [33]) where travelling cost weighs much more heavily than congestion levels and travelling times.

The study area used to calculate the scores consisted of Tlokwe municipality a service, academic, agricultural and industrial municipality with a population of 162 762 divided into 52537 households see South African Census 2011, [33] data. Tlokwe municipality is situated 120 km South West of Johannesburg in the North-West Province of South Africa.

2.3 The normalised scores of the different (criteria and factors)

In this chapter when referring to the measure scales of scores remember that different kind of data can be measured on different scales.

To summarise this consider the following measure scales for data:

- Qualitative data:
  - Zero point scale. Example: $---/0/+++$
  - Nominal (no order) scale. Example: female and male.
  - Ordinal (order) scale.
  - Binary scale. Example: yes or no.
- Quantitative data:
  - interval scale;
  - ratio scale.

Both the factor weights (see Table 2) and the factor scores (see Table 3) should be normalised such that they are dimensionless. Different methods can be used to normalise data or the

Table 3: Factors scores

Criteria and factor(s)	Normalised scores	Source of scores
1. **Density**		
1.1 S_{Gini}	0.3729	Census data
1.2 Employment-Population	0.3174	Census data
2. **Enforcement of LUT policies**		
2.1 IG_{factor}	0.25	Tlokwe LM interview
2.2 S_D	0.1	Tlokwe LM interview
3. **Accessibility**		
3.1 S_{TC}	0.25	Surveys and census
3.2 Public transport usage	0.5978	Traffic survey
3.3 d(Res,work)	0.1	Traffic survey
3.4 S_{TT}	0.5	Surveys
4. **Mobility**		
4.1 Congestion	0.5	Traffic survey
4.2 Forecasted congestion	0.25	Traffic growth rates
4.3 RS	0.1	Census

dependent variable in which case the well-known statistical Box-Cox transformation may be used. To normalise data the normalisation method used depends on the measure scales of the data.

The connection between the normalisation methods and the measure scales:

- Quantitative Measure Scales: $\frac{\text{Score}-\text{Min. Score}}{\text{Max. Score}-\text{Min. score}}$ or by division by a constant, where the constant can be the maximum score, the l^∞ norm or the l^2 norm.
- Qualitative Measure Scales: Here an applicable transformation function can be used to transform the qualitative data to the interval, $[0;1]$. For example, the Zero point scale, qualitative data $---/0/+++$ can be transformed to the respective values $0; \frac{1}{6}; \frac{2}{6}; \frac{3}{6}; \frac{4}{6}; \frac{5}{6}; 1$.

In the analysis step the different normalised factor weights are combined (using mathematical routines, which may be written into computer programmes) with the respective normalised factor scores to form a single ranking function. Furthermore, the respective ranking functions are used as the values scored by the applicable local municipality on the scorecard.

Next, consider examples of analytic methods to be used for the combination of these two components, refer to Marler et al. [1] for more technical detail:

- Linear additive method:
- Weighted Tchebycheff method;
- Weighted global criterion method;
- Weighted min-max method;
- Exponential weighted method;
- Weighted product method;
- Goal programming method;
- Bounded objective function method.

Table 4: Analysis using different methods.

Criteria and factor(s)	Linear additive method	Exponential weighted method
1. **Density**	0.0659	0.282
1.1 Gini concentration ratio	0.0373	0.153
1.2 Employment-Population	0.0286	0.129
2. **Enforcement of LUT policies**	0.0228	0.1605
2.1 Integration	0.0163	0.086
2.2 Database	0.0065	0.074
3. **Accessibility**	0.1046	0.467
3.1 Travel Cost	0.0528	0.302
3.2 Public transport usage	0.0418	0.132
3.3 d(Res,work)	0.0059	0.067
3.4 Time Travelling	0.01	0.033
4. **Mobility**	0.0973	0.465
4.1 Congestion	0.0595	0.208
4.2 Forecasted congestion	0.0295	0.161
4.3 Road safety	0.0083	0.096

- Lexicographic method.

In Table 4 only two of these analytic methods were used, namely the Linear additive method and the Exponential weighted method. With the Linear additive method it is impossible to obtain all points on the non-convex portions of the Pareto optimal set therefore, the Exponential weighted method was also required to capture points on the non-convex portions of the Pareto set.

Let k_i denote the number of applicable factors of criteria i and p be equal to the total number of criteria (as in Appendix E).

Then by using the Linear additive method the ranking of the applicable Local Municipality (i.e. the total score on the scorecard) is determine by the summation over all the criteria rankings, say R_{C_i} where

$$R_{C_i} = \sum_{j=1}^{k_i} \text{normalised weight of factor}_j * \text{normalised score of factor}_j.$$

Hence, the

$$\text{Ranking of the applicable Local Municipality} = \sum_{i=1}^{p} R_{C_i}.$$

Furthermore, for the Exponential weighted method:

$$R_{C_i} = \sum_{j=1}^{k_i} [\exp^{\text{normalised weight of factor}_j} - 1] * \exp^{\text{normalised score of factor}_j}.$$

The above analytic methods can be used to rank the different criteria, factors and local municipalities and thereby determine the weak and strong points of each local municipality.

Table 4 gives an example where these analytic methods were applied to Tlokwe Local Municipality, but this table must be extended to include comparison with local municipalities that are comparable.

3 CONCLUSIONS

The chapter illustrated the use of vector optimisation methods in land use and transportation modelling. This is supported by the development and application of a scorecard to measure the integration between land use and transportation. This is a simplified way using predetermined criteria with application of predetermined weights to describe and assess this relationship. It provides a simplified way to improve, manage and assess the interface between various municipalities on a local, regional and national level. In this manner land use and transport integration for various sizes of cities may be compared. This provides an opportunity to assess present relationships as well as to predict future relationships based on specific planning proposals. The model is very dynamic as it may be adapted through refinement of the scorecard depending on the area within a spatial system where it is applied. The framework illustrates the application of basic mathematical and modelling principles and consists of a framework where no or limited access to modelling software exists.

ACKNOWLEDGEMENT

The author gratefully acknowledges the National Research Foundation (GUN No. 81221) for financial support.

APPENDIX A: MUTUALLY DISJOINT

From any statistical textbook it follows that the matrix X is **perfectly multicollinear** if one of the columns of matrix X is an exact linear function of one or more of the other columns, i.e. the columns of X are linearly dependent. Therefore if criteria are **linearly dependent**, one criterion can be expressed as an **exact** linear function of one or even all of the other criteria. In reality every AHP matrix will contain some degree of multicollinearity because of some degree of correlation among the criteria which follows from the "consistent enough" property. Thus, we must identify whether the degree of multicollinearity in the AHP matrix is sufficiently low as to guarantee an inexact linear relationship (imperfect multicollinearity) between the criteria i.e. **mutual disjointness** between the criteria. For example an imperfect linear relationship between criterion C_1 and criterion C_2 imply there is a random non-zero error term e such that

$$C_1 = C_2 + e.$$

APPENDIX B: THE ITERATIVE ALGORITHM

This is an improvement, extension and adaptation to Matlab of the general work done by Peters et al. [34].

```
Algorithm: Analytical Hierarchy Process Matrix
        lambdaMax = eigs(A, 1);
        CRbefore = (lambdaMax − n)/(n − 1);
if n == 3 CRgoal = 0.05; else if n == 4
        CRgoal = 0.08;else if n == 5
        CRgoal = 0.1; else if n == 6 CRgoal = 0.2
        end end end end
        while CRbefore > CRgoal
                CALL function Find Greatest Deviation
                CALL function Matrix Adjustment
while (CRafter > CRbefore) AND (CRbefore > CRgoal)
                CALL function Next Greatest Deviation
                CALL function Matrix Adjustment end end
                CALL function Order of criteria (or factors) before changes
                CALL function Order of criteria (or factors) after changes
```

```
function Find Greatest Deviation
    a = size(A, 1); t = a^a + 1; cg = zeros(1, t); aijCorr = zeros(a, a); cgmax = 0;
        for i = 1 : a for j = 1 : a for k = 1 : a
    aijCorr(i, j) = A(i, k) * A(k, j);cg(h) = abs(A(i, j) − aijCorr(i, j));
        if (cg(h) == max(cg)) AND (cg(h) > cgmax)
cgmax = max(cg); maxi = i; maxk = k; maxj = j; end
                h = h + 1; end end end end
```

function Matrix Adjustment

$$b = size(A,1);\ aC = zeros(b,b);$$
$$aC(maxi,maxj) = A(maxi,maxk) * A(maxk,maxj);$$
$$cgrevise = A(maxi,maxj) - aC(maxi,maxj);$$
$AT = A$; if $cgrevise\ = 0$

if $cgrevise < 0$ AND $(A(maxi,maxj) <= 8)$

$$AT(maxi,maxj) = A(maxi,maxj) + 1;$$
$AT(maxj,maxi) = 1/(AT(maxi,maxj))$; end

if $cgrevise > 0$ AND $(A(maxi,maxj) >= 10/9)$

$$AT(maxi,maxj) = A(maxi,maxj) - 1;$$
$AT(maxj,maxi) = 1/(AT(maxi,maxj))$; end end

$$n = size(AT,1);\ NlambdaMax = eigs(AT,1);$$
$$CRafter = ((NlambdaMax - n)/(n-1));$$
if $CRafter <= CRbeforeA = AT;$

$CRbefore = CRafter$; end end

function Next Greatest Deviation

$$d = size(A,1);$$
$$e = 1;$$
$$t = d^d + 1;$$
for $m = 1 : t$

if $cg(m) == cgmax$

$$cg(m) = 0;$$
$cgmax = max(cg)$; end end

$$l = 1;$$
for $i = 1 : d$ for $j = 1 : d$ for $k = 1 : d$

if $(cg(l) == cgmax)$

$maxi = i; maxk = k; maxj = j$; end

$l = l + 1$; end end end

APPENDIX C: IMPROVED METHOD I

Improved Method I:

$$lMax = eigs(B,1);\ [VD] = eigs(B);\ eigenvector = V(:,1);$$
$$t = sum(eigenvector);\ z = eigenvector./t;\ n = size(B,2);$$
$$s = n*n;\ k = 1/(n-1)(n-2);$$
$$M = zeros(s,1);\ L = zeros(s,1);\ Y = zeros(s,1);$$
$$v = 1;\ j = 1;$$

while $v < s$

for $i = 1 : n$

for $j = 1 : n$

$$Y(v) = ((B(i,j)*z(j)*n)/lMax) - z(i);$$

$v = v + 1;$ end end end

$$L = Y.^2;\ d = length(L);\ h = d/n;$$
$$T = zeros(h,1);\ lambda = zeros(n,1);$$
$$s = 0;\ i = 1;\ p = 1;\ v = n;\ c = 1;$$

while $c <= h$

for $j = p : v$

$s = L(j) + s;$ end

$$T(i) = s$$
$$lambda(i) = sqrt(k*som);$$

$som = 0;\ i = i+1;\ p = v+1;\ v = (i*n);\ c = c+1;$ end

$$UBS = z + lambda \quad LBS = z - lambda$$

APPENDIX D: DETERMINING THE STARTING AND END PRIORITIES

```
function Order of criteria (or factors) before changes
        p = 1;
        while p <= w
          for r = 1 : n
        if sumRb(p) == sortRB(r)
        orderRb(p) = r;  end end
          p = p + 1; end
        orderRb
```

```
function Order of criteria (or factors) after changes
        p = 1;
        while p <= w
          for r = 1 : n
        if sumRa(p) == sortRa(r)
        orderRa(p) = r;  end end
          p = p + 1; end
        orderRa
```

APPENDIX E: THE EVALUATION OF THE ERROR FORMULA

a_{ij} := Denote how many times the i-th criteria (factor) is more important than the j-th criteria (factor);

m := Total number of (independent) factors;

p := Total number of (independent) criteria $\Rightarrow p < m$;

k_i := The number of factors for criteria i,

$$\Rightarrow \left[\sum_{i=1}^{p} k_i\right] = m, \quad \Rightarrow k_i < m, \forall i \in [1, p];$$

$w_i^{(j)}$:= The weight of the j-th factor of criteria i, $\quad j \in [1, k_i], i \in [1, p]$;

w_i := The weight of criteria i, $\quad \Rightarrow \sum_{j=1}^{k_i} w_i^{(j)} = w_i, \forall i \in [1, p]$.

Then $$1 := \sum_{i=1}^{p} w_i = \sum_{i=1}^{p}\sum_{j=1}^{k_i} w_i^{(j)} = \sum_{i=1}^{p}\sum_{j=1}^{m} w_i^{(j)}.$$

Now, let

$$\text{mean}(w_i) := \mu_i = \frac{1}{k_i}\sum_{j=1}^{k_i} w_i^{(j)}, \quad \forall i \in [1,p], \tag{E.2}$$

$$\text{standard deviation}(w_i) := \Delta\mu_i = \sqrt{\frac{1}{k_i - 1}\sum_{j=1}^{k_i}(w_i^{(j)} - \mu_i)^2}. \tag{E.3}$$

Now, to determine the criteria weights consider

$$\begin{aligned}
& A\underline{\mu} = \lambda_{max}\underline{\mu} \\
\Leftrightarrow \quad & A\begin{bmatrix}\mu_1 \\ \vdots \\ \mu_p\end{bmatrix} = \lambda_{max}\begin{bmatrix}\mu_1 \\ \vdots \\ \mu_p\end{bmatrix} \\
\Leftrightarrow \quad & \sum_{j=1}^{p} a_{ij}\mu_j = \lambda_{max}\mu_i, \quad \forall\, i = 1,\ldots,p \\
\Leftrightarrow \quad & \frac{\sum_{j=1}^{p} a_{ij}k_i\mu_j}{\lambda_{max}} \overset{(E.2)}{=} \sum_{j=1}^{k_i} w_i^{(j)} \quad \forall\, i \in [1,p] \\
\Leftrightarrow \quad & \frac{a_{ij}k_i\mu_j}{\lambda_{max}} = w_i^{(j)} \quad \forall\, i \in [1,p], \quad \forall\, j \in [1,k_i].
\end{aligned}$$

Therefore, eqn (E.3) can be written as

$$\Delta\mu_i = \sqrt{\frac{1}{k_i - 1}\sum_{j=1}^{k_i}\left(\frac{a_{ij}k_i\mu_j}{\lambda_{max}} - \mu_i\right)^2}, \quad \forall\, i \in [1,p]. \tag{E.4}$$

From the work of Tomashevskii (see [22]) it follows that the errors of the inconsistent pairwise comparison matrices is given by eqn (E.4).

REFERENCES

[1] Marler, R.T. & Arora, J.S., Review of Multi-Objective Optimisation Concepts and Methods for Engineering. Technical Report Number ODL-01.01, University of Iowa, Optimal Design Laboratory, Iowa City. IA, 2003.

[2] Saaty, T.L., *The Analytic Hierarchy Process*, McGraw Hill: New York, 1980.

[3] Saaty, T.L., How to make a decision: the analytic hierarchy process. *European Journal of Operational Research*, **48**(1), pp. 9–26, 1990.

[4] Saaty, T.L., *Decision Making for Leaders*, RWS Publications: Pittsburgh, PA, 2001.

[5] Saaty, T.L. & Vargas, L.G., *Models, Methods, Concepts and Applications of the Analytic Hierarchy Process*, Kluwer Academic Publishers: Boston, MA–Dordrecht–London, 2001.

[6] Vargas, L.G., An overview of the analytic hierarchy process and its applications. *European Journal of Operational Research*, **48**(1), pp. 2–8, 1990.

[7] Koski, J. & Silvennoinen, R., Norm methods and partial weighting in multicriterion optimisation of structures. *International Journal for Numerical Methods in Engineering*, **24**(6), pp. 1101–1121, 1987, DOI:10.1002/nme.1620240606.

[8] Schoeman, I.M., Determination and ranking of integration measures for land use and transportation applications, *Urban Transport XXI*. WITpress, pp. 15–25, 2015.
[9] Department for Communities and Local Government. Multi-criteria analysis: a manual. Available at www.communities.gov.uk, pp. 35–39, 2009.
[10] United Nations. *Indicators of Sustainable Development: Guidelines and Methodologies*, United Nations: New York, 2007.
[11] Niven, P.R., *Balanced Scorecard Step by Step for Government and Non-profit Agencies*, John Wiley and Sons: New York, 2003.
[12] Gini, C., *Variabilità e mutabilità* ("Variability and Mutability"), C. Cuppini: Bologna, 1912. Reprinted in *Memorie di metodologica statistica*, ed. E. Pizetti and T. Salvemini, Libreria Eredi Virgilio Veschi: Rome, 1955.
[13] Stats SA, Gender Series Volume I: Economic Empowerment, 2001-2014 Report, http://beta2.statssa.gov.za/publications/Report-03-10-04/Report-03-10-042014.pdf.
[14] Nicolai, T.W. & Nagel, K., Coupling transport and land-use: Investigating accessibility indicators for feedback from a travel to a land use model. Presented at *Latsis Symposium 2012 – 1st European Symposium on Quantitative Methods in Transportation Systems*, Lausanne, Switzerland, 2012, http://www.sustaincity.org/publications/SCCI_Coupling_Transport_and_Landuse.pdf
[15] The Department of Public Service and Administration, Republic of South Africa. Guideline: Improving geographic access to government service points. Available at www.dpsa.gov.za.
[16] Iaych, K. Alexeev, V. & Latipov, O., IRF World Road Statistics. Geneva, Switzerland: International Road Federation (IRF), 2009, www.arrivealive.co.za/documents/March%202011%20Road%20Traffic%20Report.pdf.
[17] De Jong, P., A statistical approach to Saaty's scaling method for priorities. *Journal of Mathematical Psychology*, **28**, pp. 467–478, 1984.
[18] Saaty, T.L., A scaling method for priorities in hierarchies, multiple objectives and fuzzy sets. *Journal of Mathematical Psychology*, **15(3)**, 1977.
[19] Saaty, T.L. & Vargas, L.G., Comparison of eigenvalue, logarithmic least square and least square methods in estimating ratio. *J. Math. Model.*, **5**, pp. 309–324, 1984.
[20] Saaty, T.L., Eigenvector and logarithmic least squares. *European Journal of Operational Research*, **48**, pp. 156–159, 1990.
[21] Saaty, T.L., *Theory and Applications of the Analytic Network Process*, RWS Publications, 2005.
[22] Tomashevskii, I.L., Eigenvector ranking method as a measuring tool: Formulas for errors. *European Journal of Operational Research*, **240**, pp. 774–780, 2015.
[23] Aguarón, J. & Moreno-Jiménez, J.M., The geometric consistency index: Approximated thresholds. *European Journal of Operational Research*, **147**, pp. 137–145, 2003.
[24] Crawford, G. & Williams, C., A note on the analysis of subjective judgement matrices. *Journal of Mathematical Psychology*, **29**, pp. 387–405, 1985.
[25] Tomashevskii, I.L., Geometric mean method for judgement matrices: formulas for errors. *Optimization and Control*, submitted.
[26] Bana e Costa, C.A. & Vansnick, J.-C., A critical analysis of the eigenvalue method used to derive priorities in AHP. *Eur. J. Oper. Res.*, **187**, pp. 1422–1428, 2008.
[27] Barzilai, J., Deriving weights from pairwise comparison matrices. *J. Oper. Res. Soc.*, **48**, pp. 1226–1232, 1997.
[28] Bozóki, S. & Rapcsák, T., On Saaty's and Koczkodaj's inconsistencies of pairwise comparison matrices. *J. Global Optim.*, **42**, pp. 157–175, 2008.

[29] Johnson, C., Beine, W. & Wang, T., Right-left asymmetry in an eigenvector ranking procedure. *J. Math. Psychol.*, **19**, pp. 61–64, 1979.
[30] Karapetrovic, S. & Rosenbloom, E., A quality control approach to consistency paradoxes in AHP. *Eur. J. Oper. Res.*, **119**, pp. 704–718, 1999.
[31] Kwiesielewicz, M. & van Uden, E., Inconsistent and contradictory judgements in pairwise comparison method in AHP. *Comput. Oper. Res.*, **31**, pp. 713–719, 2004.
[32] Monsuur, H., An intrinsic consistency threshold for reciprocal matrices. *Eur. J. Oper. Res.*, **96**, pp. 387–391, 1997.
[33] South African Census 2011, http://www.statssa.gov.za.
[34] Peters, M.L. & Zelewski, D., A heuristic algorithm to improve the consistency of judgements in the Analytical Hierarchy Process. Arbeitsbericht Nr. 18. Available at https://www.pim.wiwi.uni-due.de/en/forschung/publications/a-heuristic-algorithm-to-improve-the-consistency-of-judgments-in-the-analytical-hierarchy-process-a-4441/. Accessed on: 29 March 2017.

Part 2
Strategic approach and integration

TRANSPORT MODELLING RESEARCH AGENDA AND NEW DIRECTIONS: EVIDENCE FROM A DEVELOPING COUNTRY

JAMES CHAKWIZIRA
School of Environmental Sciences, University of Venda, South Africa

ABSTRACT
Transport modelling is an essential tool for informed and decisive transport planning, design and policy outcomes. Developing robust, adaptable and intelligent transport models that address daily transport challenges and matters is a pillar and catalyst for enhanced socio-economic growth and development. Noting that models should not only be developed to satisfy scientific and intellectual curiosity it is crucial that models play a pivotal role in reducing science to practice, technology innovation and patents, technology demonstrators, standards and protocols. This is necessary in adding value and ensuring that transport modelling interventions play an instrumental and steering mechanism role aimed at advancing human civilisation on this planet. The chapter conducts a rapid appraisal of transport modelling landscape in South Africa through taking a cross-sectional view of major transport themes and challenges being tackled and gauging the space and location of modelling in these initiatives. Major articles, research papers and policy documents in the public domain are analysed and transport modelling implications discussed. Overall the chapter identifies that not enough transport modelling has been applied in South Africa and there is need to do more in improving the transport modelling dimension of the transport discipline.
Keywords: transport modelling, empirical, development, South Africa.

1 INTRODUCTION

Transportation is an integral component of local, regional, national and the global economy. The growth of economies and populations normally stimulate a commensurate growth in the demand for passenger and freight transport as an example. Meeting the transport needs of growing economies and populations (e.g. energy requirements, infrastructure requirements, agriculture and food security, etc.) requires investment in new infrastructure, maintenance and rehabilitation. A failure to invest prudently in infrastructure can have far-reaching negative impacts on the economy [1]. Central to robust and intelligent responses to growing transportation challenges is the need to employ transport modelling tools and techniques to act as a sponge and hedge against making ill-conceived and ill-informed choices and decisions.

The post millennium development goals (MDGs) – the sustainable development goals for South Africa (a developing country) place sustainable development at the centre of development. Transport growth and development initiatives in seeking to be inclusive will require incorporating climate change, disaster risk, and food security, efficient energy use of resources as top human settlement intervention agenda for both urban and rural areas. Contemporary urban and rural transport developments constitute the heart and soul of a new wave of changes engulfing all spatial areas with differentiated impacts and outcomes. Consequently, in this context concerns are raised about the normative aspects of projects and programmes being rolled out. For example, the loss of natural habitat and biodiversity is a recurring thematic concern for green development activists, agriculturalists as well as rural landscape advocates. At the same time in urban areas, the spiralling and controversial emission of pollutants and release of greenhouse gases into the atmosphere remain a problem. The problems of high motorisation and an auto-mobile dependent society manifest themselves further in terms of congestion and various chapters of social exclusion. Addressing these matters is necessary as these pervasive challenges cannot be shelved. One way of confronting these matters is through investing in strategic long term modelling to enable scenario development as a way of managing the challenges. Such long term strategic modelling will enable the development of platforms that facilitate better decision making regarding the future infrastructure and land use structuring of both rural and urban settlements. The envisaged modelling footprint will not only be restricted in terms of what can be built, but can also be extended to incorporate the modelling of mitigation measures that can be adopted in seeking to make existing urban and rural systems work better in meeting society's growth and development needs.

It is encouraging to note that transport modelling in South Africa is occurring in a context of growing challenges facing both the first sphere of government and third sphere of government inclusive of local municipalities. Some of the major growth and development challenges in developing countries include the following, namely:

- A growth in the number of poor people living in both cities and rural areas;
- Supply side constraints to economic growth inhibiting capacity of institutions to respond to backlogs, deficits and shortages;
- An inefficient spatial development pattern and form (both urban and rural): The sprawling, low density nature of South Africa's cities imposes significant costs on households, firms and the public sector. This development pattern is evident in all previous Colonial countries within the developing world;
- Rising demand for urban services: Population growth and declining average household sizes place increasing pressure on municipalities, to deliver services; and
- A demanding social/elected citizenry.

Against this background, transport modelling initiatives in developing countries generally and South Africa in particular need to be both sensitive and progressive in tackling complex, dynamic and evolving transport and development challenges.

1.1 Working definition of a transport model

According to the Oxford Dictionary (1976), a (transport) model is "a simplified description of a system, etc. to assist calculations and predictions". Generally, a transport model is considered to be a simple representation of reality [2]. In simple terms, a model represents a simple theoretical abstraction of reality that represent transport systems in such a way that essential features crucial to the development of a transport theory and its application are identified and highlighted [3]. Models can take on different forms, such as maps or mental models. Various 'tools' in the form of computer programmes or software are utilised to produce models [4].

1.2 Purpose of transport models

Among numerous purposes that a transport model serves includes enabling the generation of probable scenarios through the forecasting function. Robust modelling provides evidence based facts and figures to use in deciding on alternatives as an example involving infrastructure investment; determining future transport mobilities and anticipated volumes of traffic on various sections and segments of a network etc. Consequently, modelling play a catalytic and facilitative role in enhancing and supporting the decision-making processes and systems in institutions and organisations. Invariably, transport modelling is a strategic tool for understanding transport gaps, deficits so as to allow the crafting of strategies and mechanism aimed at optimising existing infrastructure as an example.

Transportation planning, in general, assumes that the future can be predicted from an abstraction of the current realities. Undertaking such a forecasting exercise requires a clear understanding of the vision, mission, strategies, targets and objectives that an institution or organisations seeks to achieve in the future. At the same time, there is need to craft a model which can appraise the effects of different interventions on the performance of the existing or proposed systems. Consequently, transport modelling will not be enough in the absence of an appraisal framework that facilitates comparison of outcomes with operational strategies [5]. Box 1 presents famous quotes on the purpose of (transportation) models which remain critical guiding thinking and framing ideas in contemporary times.

1.3

1.3 Aim of the chapter

The chapter seeks to explore and locate the space of transportation modelling in South Africa specifically and developing countries in general. This is achieved through conducting a rapid appraisal of literature review of documents in the public domain.

1.4 Objectives of the chapter

The chapter seeks to achieve the following, namely:

1. Describe the purpose of transport modelling;
2. Explore the transportation modelling applications and implications in South Africa; and
3. Make recommendations on strengthening transportation modelling in South Africa in particular and the developing countries in general.

1.5 Research methodology

The chapter reviews literature making use of a desktop survey methodology. In addition, interviews with selected key informants and stakeholders were conducted to complement secondary data analysis. The case study adopted is South Africa while the thematic focus area is transport modelling.

Box 1: Classical quotes on the purpose of (transportation) models. *(Source: author's own compilation.)*

> *If you cannot accurately predict the future then you must flexibly be prepared to deal with various possible futures.* Edward de Bono, *Serious Creativity.*
>
> *"As simple as possible – but no simpler."* Albert Einstein (14 March 1879–18 April 1955).
>
> *"The purpose of models is not to fit the data but to sharpen the questions,"* says Van Vuren. It is attributed to Samuel Karlin (8 Jun. 1924–18 Dec. 2007), an American mathematician. Samuel Karlin Delivered at the Eleventh R.A. Fisher Memorial Lecture, Royal Society (20 April 1983). In Carl C. Gaither and Alma E. Cavazos-Gaither, *Statistically Speaking: A Dictionary of Quotations*, 1996, p. 140.

1.6 Research questions

In order to achieve the chapter's aim of exploring and locating the space of transportation modelling in South Africa specifically and developing countries in general, the following framing questions were framed to operationalise the research aim, namely:

1. What are the main contemporary transportation modelling application areas?
2. What are the main transportation modelling gaps, opportunities and challenges? and
3. What are the major intervention levers for generating the next generation of transportation modelling packages in South Africa in particular and developing countries in general?

2 LITERATURE REVIEW

Numerous integrated land-use transport models exist and are available for use by transportation, land use and engineering professionals. Making a choice on the appropriate land use transport models requires an understanding of an institutions capacity, context and culture. The bottom line relates to employing land use and transport models that maximize the goals and objectives of growth and development. While the generic framework of models is discernible in all models, a detailed review of models highlights significant differences among models. Usually the differences emanate from the following aspects, namely: comprehensiveness; model structure; theoretical origins; modelling techniques; dynamics; data requirements; calibration and validation.

2.1 Forecasting transport impacts and outcomes methods

Overall, forecasting impacts and outcomes of integrated land-use transport policies in rural and urban communities has proved a very difficult task in practice within developing countries. This is because the land use and transport sectors undergo rapid change and are impacted by different concurrent processes and changes at all scales. Despite these challenges, a threefold method of impact and outcome modelling exist, namely:

- Stated preference method: this method relies on asking people about their anticipated reaction to changes such as increased transport costs or land use restrictions for their mobility patterns, activities and behaviour.
- Revealed preference method: this method is based on drawing conclusions from empirically observed behaviour regarding how people travel, why they travel and the impacts of travel behaviour on the environment.
- Mathematical prediction transport models: this method comprises the application of mathematical models to simulate human decision making and its consequences on land use and transport decision framework.

Fig. 1 presents a schematic illustration that highlights that transport intervention impacts change over time, by place and is not evenly distributed within space. Using transport modelling to assist with redressing spatial equity, market distortions correction as well as promoting greater growth and development outcomes is central to interventions. Generally, the value of transport benefits must decline if indirect economic benefits are to grow [6].

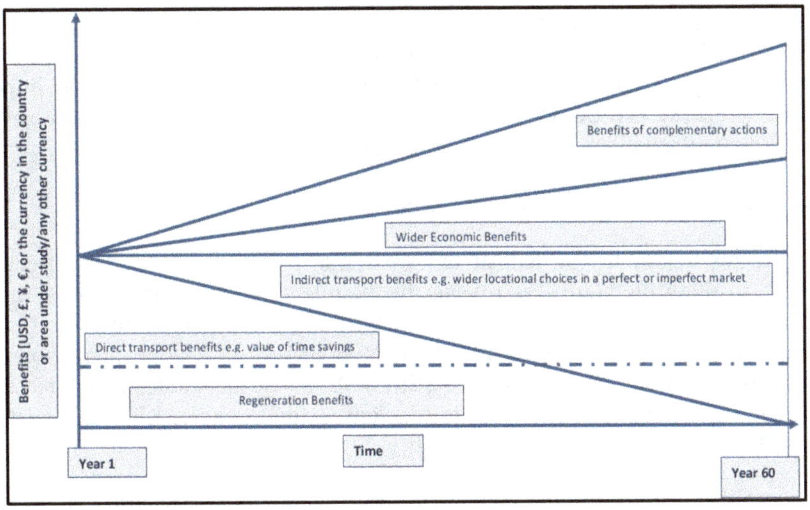

Figure 1: Schematic illustration of transport impacts time and spatial distribution. *(Source: author's own compilation.)*

2.2 Transport modelling development realities and considerations

Transport systems and outcomes can be modelled through employing an endogenous or through adopting an exogenous transport model. Generally, the rural and urban fabric (i.e. including infrastructure networks and land use patterns) undergo slow changes over-time. However, workplaces and housing sub-components have a capacity to change relatively slowly. This is in contrast to the employment and residential population sub-systems that adjust their spatial behaviour fairly quickly to changing circumstances. Freight transport or travel destinations are the most flexible phenomena of urban spatial development. This is because these can be modified almost instantly according to changes in congestion or fluctuations in demand. However, the rural and urban environment is much more complex regarding its temporal behaviour. While it is acknowledged that the land use and transport sector has benefited from a wave of cyclic improvements to modelling innovations over the past, this has not resulted in a modelling challenge free environment. As an example, the transport sub-models used in most current land-use transport models do not fully incorporate and apply state-of-the-art activity-based modelling techniques but the traditional four-step travel demand model sequence which is inadequate for modelling behavioural responses to many currently applied travel demand management policies. Micro-simulation in transport modelling provides the greatest prospect for overcoming challenges associated with the need to reproduce the complex spatial behaviour of individuals on a one-to-one basis. In addition, advances in geographical information systems and remote sensing science have led to increased spatial resolution of models. However, more transport modelling research work regarding neighbourhood (micro level) scale policies and effects is a ripe area for research. More recently Land Use and Transportation interaction (LUTI) models have been extended to deal with regional and inter-regional impacts of transport development [7], [8]. The key weaknesses of LUTI models arise from base model assumptions for example the input-output models are static in nature, dependent on existing patterns of behaviour and can only solve problem through ensuring that equilibrium is reached in each relevant market. Computable general equilibrium models (CGE models) have at their core the possibility of assuming that consumers display preferences over differentiated goods which are produced by imperfectly competitive firms [8], [9]. Because of this use of a utility function CGE models can make a direct estimate of the welfare effects resulting from a change.

2.3 Mainstream transport models and application domains

Table 1 presents a summary of mainstream urban transport models including strengths and limitations. In short, literature scan reveals the main land use and transportation modelling categories as falling under the following, namely: Land use and Transport (LUT), cellular automata, system dynamics, agent based models (ABMs) and spatial interaction gravity-based models (SE) and Economic Models (EMs). A few of the studies integrated different types of models and there may be some overlaps – for example, LUT and agent based models (ABMs – such as UrbanSim) [4], [10].

Table 1: Urban and rural subsystems represented in land-use transport models. *(Source: [11]–[26].)*

Models	**Very slow**		**Speed of change**				**Immediate**	
			Slow		**Fast**			
	Networks	**Land-use**	**Workplaces**	**Housing**	**Employment**	**Population**	**Goods transport**	**Travel**
BOYCE [Combined models of location and travel choice]	×				×	×		×
CUFM [California Urban Futures Model]	[×]	×	×	×	×	×		[×]
DELTA [Land-use/economic modelling package]	]×]	×	×	×	×	×		[×]
ILUTE [Integrated Land Use, Transportation, Environment modelling system]	×	×	×	×	×	×	×	×
IMREL [Integrated Model of Residential and Employment Location]	×	×	×	×	×	×		×
IRPUD [Model of the Dortmund region]	×	×	×	×	×	×		×
ITLUP [Integrated Transportation and Land-Use Package]	×	×			×	×		×
KIM [Non-linear urban equilibrium model]	×				×	×	×	×
LILT [Leeds Integrated Land-Use/Transport model]	×	×	×	×	×	×		×
MEPLAN [Integrated modelling package]	×	×	×	×	×	×	×	×
METROSIM [Microeconomic Land-use and Transport model]	×	×	×	×	×	×		×
MUSSA ['5-Stage Land-Use Transport Model']	[×]			×	×	×		×
PECAS [Production, Exchange and Consumption Allocation System]	×	×	×	×	×	×	×	×
POLIS [Projective Optimization Land Use Information System]	[×]	×			×	×		]×]
RURBAN [Random-Utility URBAN model]	[×[	×			×	×		[×]
STASA [Master-equation based transport and urban/regional model]	×	×	×	×	×	×	×	×
TLUMIP [Transport and Land Use Model Integration Program]	×	×	×	×	×	×	×	×
TRANUS [Transport and Land-use model]	×	×	×	×	×	×	×	×
TRESIS [Transportation and Environment Strategy Impact Simulator]	×	×	×	×	×	×		×
URBANSIM [Microeconomic model of location choice of households and firms]	[×]	×	×	×	×	×		[×]

Key: [×] Provided by linked transport model.

Overall, in South Africa and within most developing countries the high motorisation and population growth places both rural and urban settlements under different forms of stress and crisis. The need to prepare and plan adequately for rapidly urbanizing and dynamic population shifts and changes in the size, location, density and scale of people, space and institutional implications need clear unpacking. For example, in 2010 Gauteng had a motorisation level of 314 vehicles per thousand people, far above the national average and more than 3.5 times higher than the 85 vehicles/1000 of the least motorised province, Limpopo [27]. Table 2 presents motorisation trends in South Africa.

In many developing countries, South Africa included the considerable infrastructure required for a multi-modal transport system is usually taken for granted as a feature of daily life. Yet at the same time the growth of populations and economies entails that demands on transportation infrastructure and services increase over time and space bringing about new waves of development complexities that require innovative ways of management. This projected and anticipated demand on infrastructure planning and services serves as motivation for the ever increased and improved projections relating to the future demand for transport and the energy required meeting existing and future demand [1], [8], [15].

The next section explores the findings and survey results.

3 DISCUSSION OF RESULTS AND FINDINGS

3.1 Linkage between transport modelling and development in South Africa

Table 3 presents a summary of some of the South African studies examining the links between transport modelling and development. The review reveals that transport modelling opportunities and initiatives exist. These are however hamstrung by being fragmented, ad-hoc, not integrated and therefore lack interoperability. However, these scattered data hungry models represent solid departure points on which strong, robust and authentic transportation modelling platforms and a data meta-language can galvanise from.

Table 2: Motorisation of South Africa's provinces compared to GDP and land area share. *(Source: [1].)*

Province	**Population (2011)**	**Total self-propelled vehicles (Dec. 2011)**	**Share of vehicles (%)**	**Motorisation (veh/1000 persons)**	**Contribution to national GDP (2010)**	**Land area share (%)**
Eastern Cape	6,829,958	628,529	7%	92	7.7%	13.8%
Free State	2,759,644	478,546	5%	173	5.5%	10.6%
Gauteng	11,328,203	3,560,678	39%	314	33.7%	1.4%
KwaZulu-Natal	10,819,130	1,268,984	14%	117	15.8%	7.7%
Limpopo	5,554,657	474,225	5%	85	7.2%	10.3%
Mpumalanga	3,657,181	585,628	6%	160	7.0%	6.3%
Northern Cape	1,096,731	195,094	2%	178	2.3%	30.5%
North West	3,253,390	457,286	5%	141	6.7%	8.7%
Western Cape	5,287,863	1,485,018	16%	281	14.1%	10.6%
Total	**50,586,757**	**9,133,988**	**100%**	**181**	**100.0%**	**100.0%**

Table 3: Summary of South African studies examining links between transport modelling and development: some insights. *(Source: [27]–[29].)*

Identity	Geographic scale	Conclusions
Driessen Commission (1974): Report on Committee of the Enquiry into Urban Transport facilities in the Republic	National	• Transport alternative; light rail and underground (tube) transport proposals • Pre-feasibility and modelling studies to preclude implementation of such initiatives
Perold A.D. (2000): An Appropriate Strategic Modelling Approach in South Africa	National	• Strategic (macro) travel demand modelling needs in South Africa include the following, namely: **1** Interface between travel demand modelling and GIS; **2** Trip chaining (e.g. the ability to not only model a trip from home to work, but also; allow for dropping off kids at school; **3** Time of day modelling (e.g. not just the am peak hour); **4** Presentation of transportation plan results for meetings with officials, politicians, public and other interested parties; **5** Public transport routing and scheduling; **6** User-friendly simulation; **7** Modal integration; and **8** The next generation of travel demand forecast • It is practical to use simulation techniques which are sensitive to people's perceptions and opinions, and the way in which they make trip making decisions
Kane L. and Behrens R. (2002): Transport planning models – an historical and critical review	National	• Conventional Large Scale Urban Transport Models – based on a motorization society rather than an urbanizing society (including land-use transport interaction models; disaggregate methods and micro-simulation methods; dynamic methods and activity-based methods, etc.) • Aggregate four-stage models not enough and to be complemented by updating models to local conditions • An expanded approach to transport focuses on individuals instead of vehicles • lives instead of trips incorporating travel substitution, trip inducement, trip suppression etc.
Jennings L (2007): A sustainable transport model for South African cities	National	• Network of transport cities sharing ideas and best practice regarding transport modelling initiatives • A viable way of building capacity for generating locally appropriate transport solutions
DoT 2007: Rural Transport Strategy	National	• Rural Accessibility Modelling (inferred not explicitly stated) • Rural Accessibility Index since developed
Robinson A (2009): Transport modelling with reference to the Gauteng Freeway Improvement Project	Provincial	• The development of the GFIP strategic traffic model was a challenge due to the time constraints. Used previous models as building blocks (i.e. The Gauteng Transport Study of 2000 (GTS 2000), Gauteng provincial model; The Super Highways model, which was a modification of the GTS 2000 developed to test the possibility of tolling provincial roads in the province)
CSIR (2009): Car Ownership study and car use study in South Africa	National	• Car ownership policy and decision making model • Car ownership model • Public transport improvement modelling
DoT (2010): National Transportation Master Plan	National	• Different transportation scenario requirements • Transport Modelling by mode, scale and over time

Table 3: Continued.

Identity	Geographic scale	Conclusions
Wray C., Musango J. and Damon K. Gauteng City-Region Observatory (2013): Modelling urban spatial change: A review of international and South African modelling initiatives	National	• LUT models currently incorporate transportation modelling processes of trip generation, distribution, modal split and assignment explicitly and are consistent with discrete choice methods based on utility maximising, specifically in their simulation of trip-making • More recently, a new generation of micro-simulation models that attempt to account for the behaviour and interactions of individual agents in space and time have been developed • A few examples of newer LUT models include: UrbanSim (Waddell, 2002), Integrated Land Use Transportation Environment (ILUTE) (Salvani and Miller, 2005) and Integrated Land-Use Modelling and Transport System Simulation (ILUMASS)

3.2 Transport modelling projects in South Africa: some examples

A number of transport modelling projects exists in South Africa. These are found at different spheres of government as well as cut across research institutions and Universities. Despite the fact that the majority of the systems are developed as stand-alone projects and are not integrated they however constitute a good point of departure for developing future integrated transport modelling platforms in South Africa. Table 4 presents some of the specific examples of transport modelling projects in South Africa.

3.3 Transport modelling initiatives: the case of the Department of Transport

The Department of Transport is the government department which in terms of its mandate provides leadership in the promotion of sustainable transport development among other important objectives. One of the areas of leadership is transport leadership through modelling interventions to assist decision makers regarding different transport future pathways and options. Table 5 presents examples and opportunities for transport planning as informed by the Department of Transport Strategic Plan.

Table 6 presents the DoT's detailed performance in 2012/13 per programme, including some selected highlights and implications for transport modelling in South Africa.

Table 7 presents transport modelling research gaps and opportunities in South Africa as informed by this research and analysis.

Table 8 presents some existing transport modelling governance in South Africa. Developing robust transport governance frameworks and systems is central to moving ahead the transport modelling agenda in any area or region.

3.4 Key transport initiatives and modelling implications in Gauteng province, South Africa

While some of the transportation applications areas in South Africa have been identified in Tables 4–8, Table 9 presents the gaps, opportunities and challenges related to transport modelling in South Africa. In summary, Table 9 presents that the problem in South Africa is not that there is no transport modelling. It indeed exists but as pockets of scattered transport modelling initiatives that do not have high impact. There is need to upscale the transport modelling applications by scale as well as intensity. Making use of Table 2, the intention of transport initiatives in Gauteng province is to deliver an enhanced road and

Table 4: Examples of specific transport modelling projects in South Africa. *(Source: [27]–[29].)*

Identity	Description	Application	Issues
Quantifying the energy needs of the transport sector for South Africa: A bottom-up model. This was developed by the Energy Research Centre in 2012	The project focused on the development of a number of models which, when combined, can be used to develop scenarios to answer questions around the likely future energy and infrastructure requirements of the transport sector and its major influences in terms of both energy and emissions.	Passenger transport Freight transportation	• Data with which to populate transport sector models is sparse in South Africa and hence a broad range of input assumptions had to be made to facilitate finalization of the model. • Transport energy demand under business as usual will double by 2050. • The vehicle fleet will have grown to exceed 24 million road vehicles by 2050 from a baseline of 8.2 million in 2010. Of these, over 17 million will be private vehicles, excluding light commercial vehicles. • It is possible to constrain the increase in energy demand to less than 30% with a trend towards stabilisation through policy instruments. • Modelling of intracity and intercity travel within and between provinces would allow the calculation of future transport fuel use for each province, which will aid decision-making around infrastructure investment and opportunities for alternative transport modes. • Finally, collaboration between researchers in South Africa working in the transport sector would greatly enhance both base year data and transport sector modelling.
Reducing greenhouse gas emissions and facilitating the transition to a green economy. This was developed in 2013 by the National Treasury.	The National Treasury's modelling is based on a dynamic computable general equilibrium (CGE) model that seeks to capture the complex economic relationships between producers and consumers, and between domestic and foreign economies. The study models the impacts of a carbon tax imposed upstream on fossil fuel inputs (coal, crude oil and natural gas) and implemented gradually over a period of 10 years. The impacts of the carbon tax on key variables are considered over a period of 25 years, including the level of emissions reductions, overall output, employment, investment, competitiveness of different sectors, and different income groups.	Emission reductions for the transport sector Energy-efficient modes of freight and public (passenger) transport Use of alternative, cleaner fuels Safe and affordable public transportation A shift of freight from road to rail	• Emission reductions for the transport sector can be achieved by improving the availability of more energy-efficient modes of freight and public (passenger) transport, and promoting the use of alternative, cleaner fuels. • Low-income households spend a large proportion of their incomes. • Study indicate that the transition to a low-carbon, climate-resilient economy will depend on the current structure of the economy; the incentives for technical and behavioural changes, the way in which revenue is recycled, and the extent to which energy, transport, industrial and trade policies are coordinated with environmental policy. • The model suggests that a carbon tax, coupled with various revenue recycling options, will have a limited negative impact on economic growth and will assist in nudging the economy onto a more sustainable and low-carbon growth path. Evidence suggests that tax shifting (which involves taxing "bads", such as pollution and GHG emissions, and reducing taxes on "goods", such as payroll and income) could generate both environmental and employment benefits. The most positive impacts for the recycling of revenue emanate from the channelling of funds towards new productive investments in the respective sectors. There may be scope for further investment in low-carbon technologies by both the government and the private sector.

Table 4: Continued.

Identity	Description	Application	Issues
The creation and application of a national freight flow model for South Africa. This was developed in 2012 by J.H. Havenga and W.J. Pienaar	The NFFM focuses on defining surface freight transport (i.e. road and rail) based on the two definitive measures of flows, namely tons and ton-kilometres. The model is a descriptive model fitted on historic data.	Freight transport logistics	• South Africa suffered from a historical lack of freight-flow information that was detrimental to infrastructure planning, optimal network development and market structuring. • Due to the nature of the input information for road data, the model produces aggregate tonnages as well as corridor flow estimates, but not at the level of detailed origin-destination flows. • Subsequent to the development of the NFFM, a detailed origin-destination model was, however, developed for South Africa, based on the input-output model, with a 30-year forecast (Havenga et al. 2011). • The relative ease of updating of the NFFM (compared to the onerous input-output model) allows the two models to act as invaluable mutual validation tools, and the validity of both models have been established and is being tracked over time (CSIR, Imperial Logistics and University of Stellenbosch 2010).
UrbanSim. This was developed by the CSIR		Urbanisation modelling Fragmentation modelling Integration modelling Land-use and transport modelling	• The highly technical, data hungry and complex UrbanSim model represents the top-end of urban simulation. • The planned operationalisation of the Gauteng-wide GIS platform, or GeoGCR spatial database by the Gauteng Planning Commission (GPC), and establishment of a national observatory, proposed by the National Planning Commission (NPC), represents an opportunity for integrated modelling coordination and sharing of outputs across different spheres – local, regional and national (i.e. creation of an integrated modelling platform). • Careful consideration should be given to the possible overlapping roles and work of the national observatory and stepSA (Spatial Temporal Evidence for Planning South Africa) programme.

public transportation service that meet current and future transport requirements for the region. Table 9 presents a summary of the key transport initiatives in Gauteng Province including the transport modelling implications. The forecasted sum of implementing these transport modelling inspired initiatives is that the Gauteng Province in South Africa will move from a phase "of transport access and constraints gridlock loop" to one of "transport access and constraints opportunities motion and continuum".

Table 5: An example from DoT's Integrated Transport Planning programme objectives: implications for transport modelling. *(Source: [27]–[29].)*

Programme objective	Implications for transport modelling
Facilitate multi-modal transport planning on an ongoing basis by:	Multi-modal transport modelling
Developing an enabling consolidated transport planning databank by 2015.	Using the consolidated transport planning databank for scenario (projection) planning
Developing the Multi-modal Integrated Transport Planning draft Bill by 2015.	Multi-modal integrated transport planning modelling
Assist line function operations with socioeconomic analysis of the transport sector by providing economic modelling and transport data on an ongoing basis.	Transport modelling of commuter and travel patterns in Urban areas
Reduce cross border system costs as well as transit and turnaround times to enhance trade within the Southern African Development Community region and Africa as a whole by:	Cross-border logistics modelling
Propose innovative and applicable technologies for improved and environmentally sustainable transport systems through research, innovation, monitoring and evaluation on an ongoing basis.	Low carbon economy transport scenario modelling at all scales and by sectors
Ensure evidence based transport planning and policy making by conducting a household travel survey and publishing its result by the end of 2013/14 to support public transport and infrastructure investments.	Household Travel Survey Modelling

Table 6: DoT's detailed performance in 2012/13 per programme: some selected highlights and implications for transport modelling in South Africa. *(Source: [27]–[29].)*

Programme 02: Integrated Transport Planning (ITP): sub programme: research and innovation			
Key performance indicator	**Target**	**Achieved / not achieved**	**Transport modelling implications**
Develop a sustainable transport information portal, accessible at all tiers of government	A sustainable transport information portal	Achieved	Raw source data and information sets that can feed into various transport modelling initiatives
Develop a rural accessibility/multi-deprivation index	Accessibility index conducted for 12 rural districts: Alfred Nzo, Amajuba, Amatole, Chris Hani, Joe Gqabi, O.R. Tambo, uGu, uMgungundlovu, uMkhanyakude, uMzinyathi, uThukela, and Zululand	Achieved	Opportunity to engage in detailed rural transport accessibility modelling in these 12 rural districts
Research on the impact of freight accidents on South	Discussion document on the impact of freight accidents on South Africa's roads	Not achieved	Freight accidents modelling
Develop rural-specific indicators and targets to monitor freight performance within IDPs and ITPs	Roll out and coordinate to other rural provinces	Achieved	District and local level freight transport modelling
Identify and develop strategic rural freight logistics infrastructure	Roll out and coordinate to other rural provinces	Achieved	Freights logistics hub modelling

Table 7: Low-hanging existing transport modelling fruits: research agenda and gaps. *(Source: [27]–[29].)*

Current research area	Research gaps	Proposed research direction and new research focus area
BRT Rea Vaya	Integrated Ticketing System Expansion of BRT route system beyond core network	All 13 Cities where IRPTN is being implemented including Rustenburg
Gautrain rapid rail Link	Alternatives such as LRS Expansion of the RRL Feasibility of underground rail systems (tube transport)	Gauteng Region
IRTP (Integrated Rural Public Transport Planning)	Rural transport test case/pilot	At least 2 Rural District Municipalities in NW Province including all Pilot Rural Integrated Public Transport Pilot cases
Compact city models – development	Exploring on ToD; catalytic and flagships projects to fast track the transition	All least for a start all provincial and major towns in South Africa

Table 8: Some transport modelling governance initiatives in South Africa. *(Source: [27]–[29].)*

Governance initiative	Justification	Transport modelling scope
Gauteng Integrated Transport Modelling Centre (GITMC)	A response to the shortcomings of ad-hoc and fragmented transport modelling outcomes of commissioned research projects.	The CSIR, in conjunction with the UP was appointed to set up the GITMC as a multi-year project. The project is aimed at enhanced co-ordination and integration of transport modelling needs for governments within Gauteng by pooling resources and expertise.
Gauteng Integrated Transport Master	Generation of transport modelling scenarios for GITMP i.e. for 2010 base year and design years 2015, 2025 and 2037).	A total of 833 traffic zones covering Gauteng and surrounding areas have been modelled over a period of seven months. At the time of writing this chapter, there was no structured collaboration between the GITMC and ITMP25 modelling initiatives, highlighting the lack of collaboration and possible duplication of modelling efforts.
eThekwini's Accessibility Model	eThekwini uses the spatially explicit Accessibility Model to understand the future social services requirements of its housing planning projects. Based on population numbers, incomes, and age profiles, the model simulates the supply of and demand for social facilities across the municipal area. It also takes into account the existing backlog in facilities.	The Accessibility Model uses GIS and modelling to generate quantitative information. So far it has reinforced the need to cluster public investment in nodes and along corridors as well as aid decision making in housing planning exercises. Being a micro-model, it has signified a move beyond broad and conceptual spatial planning through the formulation and use of standards. Given that the model has proved to be very useful in matching the supply of and demand for social facilities, it has helped to avoid unnecessary and wasteful development in this regard.

Table 9: Sample of key transport initiatives and modelling implications in Gauteng Province. *(Source: [27]–[29].)*

Key transport initiative	Implementation levers	Expected outputs	Transport modelling implications
Transport Authority for Gauteng	Institutional	• Agreement on an appropriate institutional structure • Business plan and funding arrangements • Institutional development and strengthening	• Transport governance and modelling
Effective Management of Existing Transport Infrastructure	Institutional	• Gauteng Integrated Investment Road and Transport Infrastructure Plan • Utilisation and updating of The Gauteng Pavement Management System (GPMS) for road maintenance, rehabilitation and construction • Capacity building and training to establish a critical mass of experts	• Infrastructure Rehabilitation Modelling • Infrastructure maintenance Modelling • New Infrastructure Construction Modelling
Regulation and enforcement	Institutional	• Strengthening municipal traffic departments • Road Safety Action Plan • Improved over-load control and enforcement • Traffic Route Integrity Audits • CCTV along major activity corridors	• Traffic accidents modelling • Transport disaster risk modelling
Provincial-wide public transport information centre	Intelligent traffic (transport) information systems	• Common GIS, data structure and databases platforms • Link to advanced public transport management systems (APTMS), Fleet management, GPS and vehicle tracking • Scoping of platform(s), setting up call centre and supporting back office	• Single transportation planning modelling platform
"One Province One Ticket"	Transport integration	• Implementation of the Integrated Fare Management (IFM) framework and strategy • Operational integration of public transport services • Common data structures, databases and warehousing	• Integrated ticketing modelling • Price/fare costing modelling • Transport operations/costing modelling
Integration with the Commuter Rail Corridor Modernisation Project of PRASA	Transport integration	• Inclusive joint planning which does not exclude any stakeholder • Agreement on priority focus stations (i.e. Germinston, Vereeniging, Roodepoort and Leralla) • Operational integration with other modes • Non-motorised transport facilities and directional signage	• Rail freight modelling • Rail passenger modelling

Table 9: Continued.

Key transport initiative	Implementation levers	Expected outputs	Transport modelling implications
Transformation of taxi industry	Institutional	• Taxi industry re-organisation and corporatization strategy • Streamlined registration and licensing system • Business opportunities on IPTN with associated financial support • Pilot project	• Taxi corporatization modelling • Taxi operational financial modelling • Taxi route operations modelling
Access to major freight nodes	Institutional	• Integration with PRASA Passenger Rail Plan • Finalization of Integrated Public Transport Network Plan • Upgrade and development of key road links and section • Dedicated freight routes on the periphery of urban area (for freight to flow directly to and from the N1/N12 via the freight terminals to Limpopo, Mpumalanga and KwaZulu-Natal (KZN)	• Road freight modelling • Maritime transport modelling • Air freight (aviation) modelling • Freight logistics border delays modelling • Inland dry ports modelling
Restructured subsidized road-based public transport	Transport funding	• Appropriate modes to be deployed on the IPTN corridors and links (i.e. BRT, express bus, bus and mini-bus) • Integration of funding strategies • Establishment of a common basis for contracting public transport operations	• Subsidy and scenario modelling • IPTN modelling • Integrated rural public transport networking modelling (pilots)
Green public transport vehicle technologies	Transport sustainability	• Green re-fleeting strategy • Co-ordinated and planned demonstration projects including bio-fuel development and processing facilities; substantial distillery ethanol manufacture plant and; conversion of public transport vehicles	• Low carbon economy modelling and scenario building
Travel demand management, less congestion and shorter travel times	Land use and transport planning integration	• Creation of capacity to manage TDM and transport systems management programmes • Set new standards transportation infrastructure and services • Effective management of transport infrastructure • Optimisation through deployment of ITS and TDM measures • Restructuring of land use and development patterns	• Land use modelling • Transport demand modelling • Transport supply modelling

Table 9: Continued.

Key transport initiative	**Implementation levers**	**Expected outputs**	**Transport modelling implications**
International and city airports	Institutional and transport integration	• Coordination between all role-players and integration of planning • Assess impacts on land and associated development • Transport accessibility and linkages • Early anticipation and prevention of bottlenecks developing in relation to air passenger and freight movements	• Aerotropolis modelling
Pedestrian paths and cycle ways	Transport integration	• Provision of NMT infrastructure and facilities (with "urban" greening) associated with key PT nodes and Provincial roads • Finalisation of the Provincial NMT policy and strategy • 50 km pedestrian paths	• Cycling bus modelling • Walking bus modelling • NMT modelling
Continued provincial-wide mobility	Transport integration	• Utilising Gauteng Freeway network as the "backbone mobility road network" of the province, implement NATMAP strategic projects for Gauteng, namely: • Strategy 1: Develop a link between Johannesburg and Rustenburg via Lanseria Airport and the N4 Platinum Corridor; • Strategy 2: Provide a link between Sandton and the N4 Platinum Corridor; and • Strategy 3: Link the N3 to the N4 Maputo Corridor through Ekurhuleni, specifically catering for freight haulage	• Public transport modelling (bus, taxi, rail etc.) • Transport corridor modelling • Transport regional development modelling
Accessible transport	Transport integration	• Providing accessible transport for people with disabilities, persons with disabilities to have equal access to transport facilities and services available to the general public, to the greatest extent possible, in rural and urban areas • Continued improvement and development of IRPT systems namely: • BRT Rea Vaya • BRT Are Yeng • Ekurhuleni IRPTN (Aerobus?) • Gautrain rapid rail link expansion • Metro rail modernisation and expansion • Gauteng Freeway Improvement Plan and e-tolling	• Accessibility modelling • Multi-modal modelling • Inter-city movements modelling • Intra-city movements modelling

Table 2 overall resonates in the lack or existence of inadequate integrated transport modelling systems in Gauteng province in particular and developing countries transport sectors in general. It is therefore little surprise that the Gauteng Department of Roads and Transport (GDRT) has identified the need for integrated land use and transportation planning and integrated planning tools to inform robust decision- making. A recent report on modelling urban spatial change highlighted the fragmented approach to transport modelling within Gauteng [4]. Traditionally, GDRT has commissioned provincial transport modelling from transport consultants. There has also been a varied response to transport modelling at a local government level with a combination of outsourced and internal models and a range of proprietary software. The result has been a disjointed approach, limited modelling capacity and skills development within government, and a mismatch of macro and micro-scale modelling [28], [29]. In 2012, terms of reference were issued by GDRT for the establishment of the Gauteng Integrated Transport Modelling Centre (GITMC). Its aim is to create a fully operational modelling centre, co-owned, co-funded and co- staffed by a range of provincial and municipal departments. It will provide high-end planning and modelling skills for a variety of macro and micro-modelling support services to role-players across the province [1], [8], [16], [28], [29]. The Council for Scientific and Industrial Research (CSIR), in conjunction with the University of Pretoria, have been appointed to set up the GITMC as a multi-year project. The centre will initially build on the data and models generated by the CSIR UrbanSim and MATSim software urban simulation project. It will formulate urban growth scenarios at a local and provincial level taking into account current and planned transport infrastructure initiatives. The centre is planned to be fully operational by 2014 [2], [28], [29].

4 LESSON LEARNED RELATED TO RSA AND APPROACHES IN DEVELOPING COUNTRIES

The main constraints and challenges that are related to upscaling transport modelling relate to the need to address various delivery and performance bottlenecks that lead to ineffective and inadequate transport solutions. These can be summarized precisely as follows, namely:

- Adequate financial provisioning to meet research and development (R&D) transport modelling projects and interventions requirements: Financial resources remain one of the main constraints in Metropolitan and District Municipalities to fulfil their obligations in terms of transport interventions. The problem can be appreciated better if one notes that in South Africa investment in road infrastructure has on average decreased by 1% annually since 1975 [30]. Financial constraints surrounding the maintenance and operation of ambient monitoring stations; conducting oversight studies as well as implementing a package of integrated transport planning and spatial turn measures is hamstrung by lack limited resources. Financial resources are essential to ensure the successful implementation and enforcement of sustainable transport agenda.
- Limited resources of transport and spatial planning modelling experts: A solid base of critical and adequately skilled transport and spatial modelling experts that can provide sustainable and decisive leadership to turn-around transport measures and actions through projects that transform the transportation landscape. While currently the metros make use of consultants to fill the gap the problem is that the inputs are ad-hoc and difficult to integrate. For example modelling systems and bases used by different consultants are usually not compatible. Building the next generation of

transport and spatial experts to take the urban and rural regions into smart future regions requires investment in manpower development and training.

- Dynamic transport authority and progressive transportation leadership: Inter-governmental and inter-departmental communication and cooperation are major constraints in ensuring successful transport measures and actions. Without the mutual understanding between related departments such as land-use planning, road infrastructure development, urban zoning and air quality management, transport and land use commuting orientated intervention and strategies are setup to fail at the implementation phase. This also applies to the inter-governmental linkages between municipal and provincial government where synergy is essential. A holistic approach which goes over the constraints of the "silo based approaches" despite the existence of the municipal inter – structures coordinating act (i.e. in South Africa for example and similar legislation in developing countries) is therefore needed in the transport and spatial related growth and development management strategies.

Consequent to this chapter's findings, Fig. 2 presents future transport modelling thematic areas in South Africa: some prospecting thoughts. Fig. 2 shows that a five-point transport modelling research agenda is required to tackle transport modelling challenges in South Africa. A similar strategy can also be developed in other developing countries but modified and informed by the transportation growth and development context of the specific country. These five critical intervention areas are namely: travel demand and behaviour; delivery of transportation services; transportation finance; sustainable development and advanced transportation system operations and performance. If sustainable and high impact outcomes are to be realised in the transportation sector equal and simultaneous investment in all the five identified areas is critical.

Table 10 presents an elaborated version of the possible research focus areas for transport modelling in South Africa and developing countries including a clear description of the problem statement and potential projects and themes for investigation.

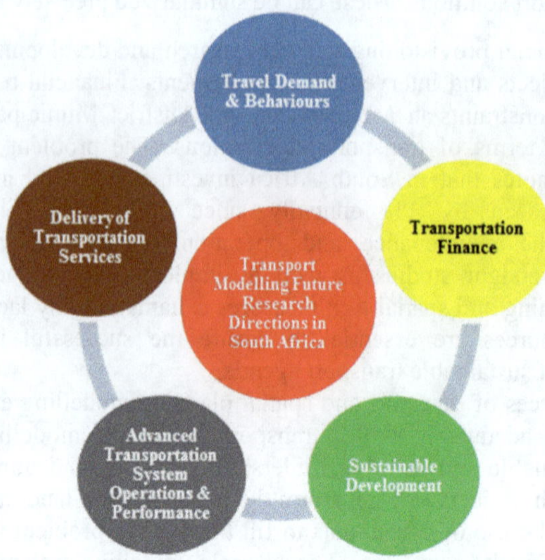

Figure 2: Future transport modelling thematic areas in South Africa: some prospecting thoughts. *(Source: author's own compilation.)*

Table 10: Travel demand and behaviours research focus area. *(Source: [27]–[29].)*

Research focus area	***Problem statement***	***Recommended potential projects/topics***
Dynamic and future travel behaviour and patterns mobilities research focus area		
Understanding fundamental ways in which travel behaviours and patterns may change in the future, and what these changes may mean in terms of transportation system needs and requirements.	Examine the long-term interplay between changing demographics, lifestyle choices, affluence/wealth, and settlement patterns on personal transportation demand. Develop improved tools for scenario analysis and forecasting that account for these considerations.	Analyse shifting patterns of population growth and personal income and its effect on transportation demand/needs. Examine a sample of existing long-range transportation and land use plans from metropolitan areas to assess and model to what extent they account for potential changes in development patterns and travel demands in response to changing demographics, technology, climate change, and other factors. Develop improved travel demand forecasting models or approaches to modelling, which account for the potential influence of long-range changes in factors affecting travel demand; this may include a structure for developing regional scenarios, accounting for risks and uncertainties regarding technologies and other factors.
Advanced transportation system operations and performance research focus area		
Addressing the potential of new technologies and approaches to create a more efficient and effectively performing transportation system.	Study the range of multi/inter-modal connectivity options, focusing on how to optimize different forms of system capacity, improve reliability, and create a seamless transportation system.	Conduct a transport modelling fore-sighting study that clearly defines the likely technological characteristics of desired future multi-modal transportation systems in order to provide a framework for areas where technological improvements could be developed by a non-transportation entity. Explore the generation of an adaptive modelling system involving all players in transportation (pedestrians, business, shippers, system operations, police, etc.) in order to facilitate intelligent and adequate designs for future systems.
Sustainable transportation research focus area		
Exploring the potential for new approaches to developing and preserving transportation infrastructure in light of potential changes in resource availability and cost.	Examine new approaches and technologies to enhance system preservation, maintenance, and renewal in light of expectations about decreasing supplies of natural resources, increased global competition for resources, and a focus on environmental sustainability.	Explore the potential consequences – intended and unintended – of developing new fuel/energy sources for transportation. This may include changes in international trade and fuel shipments (e.g., shipment of hydrogen), development of new South African-based or foreign technologies and associated impacts on the economy (e.g., new manufacturing, shipment, and energy distribution facilities), and impacts on agriculture (e.g. food production to fuel production) and other sectors of the economy. Develop a model to forecast carbon footprint and analyse potential benefits of alternative freight systems, considering modal alternative, freight bottlenecks, modal connectors, reduction in truck idling, truck – only lanes and fair pricing, rail, and water.

Table 10: Continued.

Research focus area	***Problem statement***	***Recommended potential projects/topics***
Transportation finance research focus area		
Addressing immediate funding and financing issues to address broader, longer-term approaches, which may rely on new technologies and institutional approaches.	Investigate and explore potential new financing mechanisms and approaches, considering new technologies, privatization, and institutional approaches.	Consider the future of government finance and the likelihood of privatization for parts of the transportation sector. Research legal and funding options to foster and allow for lower cost implementation of transportation projects while upholding environmental standards and requirements. Assess the ability of multi and intermodal approaches ability to minimize the effects of reduced funding.
Delivery of transportation services research focus area		
Assessing the capability to deliver transportation projects, operate services, and preserve system infrastructure effectively is essential to transport authority' success.	The knowledge economy/society plus technological changes are dramatically altering public interaction in general, with public transport authorities, and regarding transportation projects and other public goods.	Investigate the need for new public communication and involvement techniques, factoring in/capitalizing on current and anticipated social changes. Identify future increased expectations for greater transportation information from younger, more technologically sophisticated people. Assess and model future organizational structures and capacities to manage the scale of impacts and benefits of transportation. Research policy architectures to enable effective decision-making at various levels of government and across a broad spectrum of functions (e.g., transportation, environment, land-use). Identify how regulations will be categorized: life cycle costing, performance-based, or outcome-based. Identify the long-term transportation work-force composition, retention, and preparation. Consider changes in retirement patterns – specifically that people are choosing to work further into their life. Analyse role of academia in providing for the transportation workforce's future needs, both with degree programs and continuing education.

5 CONCLUSIONS

This chapter has identified that not enough transport modelling has been applied in the South Africa. The same conclusion applies to other developing countries. Scope and room for improved transport modelling in the transport discipline exist. It is also interesting to note that transport modelling main applications areas has been metropolitan and urban areas with lack of equal emphasis in small and medium sized rural towns. This gap is mirrored too in terms of greater scale of modelling initiatives in urban transport planning compared to rural transport. Looking into the future, the need to balance transport modelling applications in urban and rural areas is central to facilitating enhanced transport improvement outcomes aimed at increased balanced development, tackling regional economic disparities and inequalities as well as promoting sustainable inclusive development goals and outcomes. However, in the final analysis transport modelling alone should and cannot be expected to be the one and only ultimate answer to urban and rural transport growth and development challenges. Transport modelling interventions need to be complemented by appropriate transport policy and a legislative framework. This should be directed in the context of the correct set of transport planning, engineering and management adaptive governance systems. While getting the "transport wheels sector oiled and moving" is central to transport contribution to the economy, it is essential that similar and appropriate interventions in the other growth sectors of the economy such as health, education, mining, agriculture and manufacturing industry to name a few is correctly positioned and aligned to promote shared growth and development outcomes of a country. While encouraging growth and development in economies is always challenging, it is acknowledged that the establishment of intelligent transport and non-transport modelling platforms constitutes a critical and vital cog of development that should not be missed or easily forgotten.

In the context of South Africa in particular and developing countries, there is still a considerable need for further transport modelling work i.e. covering the full transport gamut including urban and rural road, rail, air, inland water and maritime sectors. The need to balance macro, meso with detailed and focused micro transport modelling studies and projects tailored to respond to specific changes in order to have in-depth understanding of the process of decision making in response to changed transport provision by both individuals and households, and firms is fundamental. The general observation that transport modelling tool and technique been under-utilised in South Africa and developing countries in general has resonates in the need to revamp and restructure transport studies and engineering curricula to increase the content and examples of transport modelling applications and vale add to development. In addition, research, science and funding agencies should also provide more opportunities for increased research funding with a particular bias towards promoting transport modelling projects.

While numerous studies exist that indicate the low visibility of transport modelling activities, it is encouraging to realise that pockets of excellence in transport modelling exist in South Africa and by extension in developing countries. These initiatives and projects can act as a departure point for increased transport modelling collaboration and partnerships that can yield a critical mass of transport modelling research output and technology packages aimed at tackling transport growth and development contemporary matters. Consequently, the need and requirement to establish a platform for transport modelling exchange and network in South Africa, Southern Africa and by extension developing countries cannot be over-emphasised. This envisaged Forum can act as a platform for greater dissemination and exchange of information revolving around transport and land use modelling, spatial interaction modelling systems as well as traffic and transport engineering modelling systems. A South Africa and or Southern African Transport Observatory is one way of tackling this matter. This can be tasked with the mandate to develop a critical mass of transport modelling experts and database in South Africa in particular and developing countries in general for use in addressing critical land use, infrastructure, transport and engineering growth and development issues. In addition, scope, feasibility and advisability of developing a post-graduate qualification with specialization in transport modelling as a niche area requires keen investigation. Seeking to craft and roll-out innovative ways of upscaling and replicating transport modelling patents, technology packages, protocols and procedures as well as engaging in pilot and demonstration projects should be based on the need for fostering continued dialogue and discussion in the transportation and allied sectors.

Overall, a good base for articulating transport modelling exists in South Africa although the challenges of sparse resources need to be addressed. Scattered and incompatible databanks and datasets has to be addressed. Information that is available relates to National Association of Automobile Manufacturers of South Africa (NAAMSA) Vehicle sales, The National Traffic Information System (eNATIS) vehicle population, National Transport Master Plan/ State of Logistics Survey (NATMAP/SOL) passenger per kilometre and tonne-kilometre data as well as South African Petroleum Industry Association (SAPIA/EB) fuel sales. However, despite protocol access and permission restriction to accessing such databases, with proper requests the data can be accessed and made available for research and development purposes. However, for transport modelling this data requires transformation into data sets that respond to answering questions on namely: inter-city rail commuter and freight tonnages moved; scrap rates; occupancy; load factors; road-rail share; average age and annual mileage. This is essential in making various calculations and scenarios regarding the transport sector. Once this kind of data has been gathered and analysed it is possible to engage in mathematical modelling regarding detailed vehicle passenger by mode of transport; passenger-annual kilometres; tonnes-kilometres per mode share; fuel consumption as well as total vehicle kilometres covered annually, by area etc. Conclusively therefore,

If you cannot accurately predict the future then you must flexibly be prepared to deal with various possible futures. Edward de Bono, Serious Creativity.

REFERENCES

[1] Merven, B., Stone, A. & Hughes, A., *Long term energy needs of the South African transport sector – part 1.* Working Paper, Energy Research Centre, University of Cape Town, 2012.

[2] Goyns, P., *Modelling real-world driving, fuel consumption and emissions of passenger vehicles: a case study in Johannesburg,* Thesis submitted in fulfillment of the Requirements for the degree Doctor Philosophiæ in Energy Studies, Department of Geography, Environmental Management and Energy Studies, Faculty of Science, University of Johannesburg, 2008.

[3] Batty, M., Fifty years of urban modelling: macro-statics to micro-dynamics. *The Dynamics of Complex Urban System*, eds S. Albeverio, D. Andrey, P. Giordano & A. Vancheri, Physica-Verlag: Heidelberg, pp. 1–20, 2008.

[4] Wray, C., Musango, J. & Damon, K., *Modelling Urban Spatial Change: A Review of International and South African Modelling Initiatives*, GCRO, Gauteng, 2013.

[5] Jonsson, R.D., *Analysing sustainable urban transport and land-use modelling tools and appraisal frameworks*, Division of Transport and Location Analysis, Department of Transport and Economics, Royal Institute of Technology SE-100 44: Stockholm, Sweden, 2008.

[6] Oosterhaven, J. & Elhorst, J.P., Indirect economic benefits of transport infrastructure investments. *Across the Border: Building on a Quarter Century of Transport Research in the Benelux*, eds W. Dullaert, B. Jourquin & J.B. Polak, De Boeck: Antwerp, 2003.

[7] Wegener, M. & Bökemann, D., *The SASI Model: Model Structure, SASI Deliverable 8*. Berichte aus dem Institut für Raumplanung 40, Institut für Raumplanung, Universität Dortmund: Dortmund: 1998.

[8] Bröcker, J., et al. *Territorial Impact of EU Transport and TEN Policies*, Final Report of Action 2.1.1. of the European Spatial Planning Observation Network ESPON 2006, Institut für Regionalforschung, Christian-Albrechts-Universität: Kiel, 2004.

[9] Barboux, C. & Collet, C., Spatial change analysis, 2012, http://gitta.info/SpatChangeAna/en/text/SpatChangeAna.pdf. Accessed on: 15 Nov. 2014.

[10] Iacono, M., Levinson, D. & El-Geneidy, A., Models of transportation and land use change: A guide to the territory. *Journal of Planning Literature*, **22**, pp. 323–340, 2008.

[11] Landis, J.D. & Zhang, M., The second generation of the California urban futures model. Part 2: Specification and calibration results of the land-use change submodel. *Environment and Planning B: Planning and Design*, **25**, pp. 795–824, 1998.

[12] Simmonds, D., The objectives and design of a new land-use modelling package: DELTA. *Regional Science in Business*, eds G.P. Clarke & M. Madden, Springer: Berlin/Heidelberg, pp. 159–188, 2001.

[13] Anderstig, C. & Mattsson, L.-G., Modelling land-use and transport interaction: Evaluations and policy analysis. *Network Infrastructure and the Urban Environment: Recent Advances in Land-Use/Transportation Modelling*, eds L. Lundqvist, L.-G. Mattsson & T.J. Kim, Springer: Berlin/Heidelberg/New York, pp. 308–328, 1998.

[14] Putman, S.H., Results from implementation of integrated transportation and land use models in metropolitan regions. *Network Infrastructure and the Urban Environment: Recent Advances in Land-Use/Transportation Modelling*, eds L. Lundqvist, L.-G. Mattsson & T.J. Kim, Springer: Berlin/Heidelberg/New York, pp. 268–287, 1998.

[15] Rho, J.H. & Kim, T.J., Solving a three-dimensional urban activity model of land use intensity and transport congestion. *Journal of Regional Science*, **29**, pp. 595–613, 1989.

[16] Mackett, R.L., LILT and MEPLAN: A comparative analysis of land-use and transport. *Policies for Leeds Transport Reviews*, **11**, pp. 131–154, 1991.

[17] Hunt, J.D., Calibrating the Naples land-use and transport model. *Environment and Planning B: Planning and Design*, **21**, pp. 569–590, 1994.

[18] Bröcker, J. Computable general equilibrium analysis in transportation economics. *Handbook of Transport Geography and Spatial Systems: Handbooks in Transport Volume 5*, eds D.A. Hensher, K.J. Button, K. Haynes & P. Stopher, Elsevier: Oxford, 2004.

[19] Anas, A. *NYMTC Transportation Models and Data Initiative: The NYMTC Land Use Model*, Alex Anas & Associates: Williamsville, NY, 1998.

[20] Martinez, F.J., *MUSSA: A Land Use Model for Santiago City Department of Civil Engineering*, University of Chile: Santiago, 1997.

[21] Hunt, J.D. & Abraham, J.E., Design and application of the PECAS land use modelling system. *Paper presented at the 8th International Conference on Computers in Urban Panning and Urban Management*, Sendai, Japan, 2003.

[22] Caindec, E.K. & Prastacos, P., A description of POLIS. The Projective Optimization Land Use Information System Working Paper 95-1, Association of Bay Area Governments, Oakland, 1995.

[23] Hensher, D. & Ton, T., TRESIS: A transportation, land use and environmental strategy impact simulator for urban areas. *Paper presented at the 8th World Conference on Transport Research*, Seoul, 2001.

[24] Waddell, P., Integrated land use and transportation planning and modelling: Addressing challenges in research and practice. *Transport Reviews*, **31**, pp. 209–229, 2011.

[25] eNaTiS, Vehicle population statistics for December/January 2012. Récupéré sur National Traffic Information System, 2012.

[26] Turok, I., Transforming South Africa's divided cities: Can devolution help? *International Planning 9 Studies*, **18**(2), pp. 168–187, 2013.

[27] Council for Scientific and Industrial Research Council, Gauteng Integrated Transport Modelling Centre: Draft institutional framework, 12th June 2012, CSIR Built Environment, 2012.

[28] DoT, National Transport Master Plan (NATMAP) 2050 Modelling Report, Department of Transport, Republic of South Africa: Pretoria, 2009.

[29] Shoko, M. & Smit, J., Use of agent-based modelling to investigate the dynamics of slum growth. *South African Journal of Geomatics*, **2**, pp. 54–67, 2013.

[30] Council for Scientific and Industrial Research (CSIR), Urban growth simulation with UrbanSim MATSim. Presentation at GITMC workshop. 22 March 2012.

ALIGNMENT OF SPATIAL, TRANSPORTATION AND ENVIRONMENTAL PLANNING: PRACTICES UNDERPINNING STRATEGIC APPROACHES IN DEVELOPING COUNTRIES

ILSE MARIA SCHOEMAN & CAREL BENJAMIN SCHOEMAN
North West University, South Africa

ABSTRACT

The debate and discourse for the need to align spatial planning, transportation planning and environmental management strategically, functionally and operationally to inform modelling is ongoing internationally since the early 2000s. This incorporates the articulation of the planning instruments used by the professionals within these functional fields and the way in which it is coordinated and applied as to enhance strategic planning and decision making.

With the approval of the Spatial Planning and Land Use Management Act (SPLUMA) (Act 16 of 2013) and the SPLUMA Regulations (23 March 2015) in South Africa, the last bastion of 'Apartheid' planning legislation from the previous political dispensation was reformed. Although this process is still legislatively being finalized at provincial and municipal levels of government through adapting new transformation structures, guidelines, policies and regulations, the opportunity exists to assess the impacts of alignment to improve strategic planning, modelling, coordination and decision making.

In this chapter the alignment between the policy and legislative frameworks, guidelines, processes and the need for integration will be discussed. The approach presented in this paper will especially guide planning within developing countries based on the learning experiences from spatial systems in developed countries.

Keywords: integrated planning, spatial planning, transportation planning, environmental management, transformation processes.

1 INTRODUCTION TO ALIGNMENT AND INTEGRATION

Internationally, spatial planning and its alignment and interface with transportation planning and environmental management form the focal point in attaining and promoting sustainable planning, development and delivery. The issue of misalignment between various levels of government in developing countries and more specifically South Africa dates back to 1910 during the formation of the Union of South Africa from its colonial inheritance. However, since democratization in 1994 and the promulgation of the Constitution of South Africa (Act 108 of 1996) [1], resulting in the establishment of three distinct spheres of government, the lack of alignment and integration became more prevalent.

In order to understand the context of this chapter, one should consider what is implied by the concept of *'alignment'*? From a governmental and planning perspective, '*alignment*' includes the articulation and optimization of processes contained in goals and objectives (inclusive of communication, involvement, engagement, support and capacity-building activities) to ensure efficient and effective outcomes through optimal application of planning instruments in plan formulation and decision making, as provided for in different policies and legislation across all levels of government.

'*Integration*' thus refers to processes and methodological approaches and procedures followed in planning processes through application of specific instruments and/or planning tools. *Alignment* and *integration* are often used inter-changeably. Modelling to be effective should thus be applied to alternative focuses such as spatial planning, transportation planning and environmental management.

The dilemma of alignment and integration, or the lack thereof, was also experienced internationally in planning processes in developed countries. The best example from international practice relates to the approach to assess integration in Europe of policy and research. Geerlings and Stead [2] state that 'policy' integration should include the following dynamic focuses:

- Vertical integration between different levels of government.
- Horizontal integration between sectors or professions in one organization.
- Inter-territorial integration between neighbouring authorities or authorities with a shared interest in infrastructure or resources.
- Intra-sectoral integration between different sections or professions.

Stead et al. [3] identify the following hierarchical foci in dealing with an approach towards integration: co-operation at the lowest level consisting of dialogue and information; co-ordination, coherence and consistency with emphasis on transparency; integration and joined-up policy approaches inclusive of dialogue and avoiding of conflicts. These are supported by inter-organizational coordination; inter-organizational collaboration; intergovernmental management and network management.

Essentially, many of these concepts are present in strategic planning instruments in South Africa such as the Integrated Development Plan (IDP) and environmental management processes such as Strategic Environmental Assessment (SEA) and Environmental Management Frameworks (EMFs) and related practices and planning processes. The realization of optimal alignment and integration approaches between strategic and planning instruments (in developing as well as developed countries) represents a main challenge. May et al. [4] also addressed the issues of integration of policy instruments as a challenge within urban transport policy.

In South Africa, the White Paper on Spatial Planning and Land Use Management [5] states that land-use and development decisions must promote a harmonious relationship between the built and the natural environment while ensuring that land development is sustainable over a longer term. Claasens [6] was one of the first planners to confront the interface between spatial planning, development planning and conservation in promoting sustainable development. Sowman and Brown [7] developed a framework to assist integration between the IDP process and environmental management in order to promote sustainability. Berrisford and Kihato [8] concluded that defining the relationship between planning and the environmental laws is a question that is persistently dogged by planning and environmental practitioners in South Africa. This view is derived from research by Sowman [9] and Todes [10] for a spatial system being classified in a developing country.

Todes et al. [11] address the relationship between planning and environmental management in South Africa based on a case study for the provincial spatial system of KwaZulu-Natal. From the outcome, specific research elements inclusive of purpose, method and evidence base are identified to demonstrate the relationship between planning and environmental management across spheres of government, tools applied and types of plans. Of importance to this chapter is the conclusion that the relationships between planning and environmental management in South Africa are complex and also the result of formal systems in planning and environment inclusive of elements such as agency, power practice and discourse. This describes the dynamics within developing countries and/or spatial systems accurately.

Todes et al. [11] isolate the following core components in assessing integration:

- The purposes of planning.
- Tools and mechanisms applied.
- Distribution of responsibilities.
- Resources and capacity.
- Networks, capacity and discourse.

In this regard, the work by Schoeman [12] on the interface between urban and regional planning, environmental management and transportation planning should also be noted. The deduction by Todes et al. [11] is based on the work developed by Vigar et al. [13]. Todes et al. [11] conclude, in general, that there is a need for greater integration between planning and environmental management in context with the assessment components identified earlier. It is concluded that integration is more than a technical process. At the same time, the legal system, forms of planning and environmental management and practical considerations are all fulfilling an important role in promoting inclusivity in integration. This may be attained through merging of functions; incorporation of foci, institutional integration, and coordination of vertical and horizontal linkages.

Retief and Rossouw [14] also address the theme of alignment and integration from a strategic perspective and conclude that the development of urban environmental policy is directly linked to the IDP process. According to them, the focus is to integrate sustainability into the IDP process and there is a clear convergence between legal and policy requirements for planning and environmental management present. Du Plessis [15] argues that, despite the recognition of the need for greater alignment of sustainability criteria at all levels of integrated spatial planning, some challenges and shortcomings still remain. This includes the lack of integration of sustainability principles inclusive of the application and/or use of environmental information in IDP and related processes.

With the exception of Geerlings and Stead [2], Stead et al. [3] and May et al. [4], the shortcoming in the abovementioned publications relates to the fact that it only focus on integration between spatial planning and the environment. Waddell [16] broadened the approach towards urban development for land use, transportation and environmental planning. The other authors as discussed, based their focuses on strategic spatial planning based on development instruments (IDP process) [17] and environmental management as included in Retief and Rossouw [14] and Du Plessis [15]. In order for *alignment* and *integration* to be inclusive, one should also address the inclusion of transport planning in the process of alignment and integration.

This challenge can in part be addressed by assessing the existing policy and legislative framework and, more specifically, the opportunities created by the Spatial Planning and Land Use Management Act (SPLUMA) [18] and the SPLUMA Regulations [19]. Several policy documents such as the National Transport Master Plan (NATMAP) [20]; the National Development Plan (NDP) [21]; the Draft Integrated Urban Development Framework (DIUDF) [22], and the Draft National Land Transport Strategic Framework (NLTSF) [23] recognize the need for alignment and strategic integration between focuses as dealt with in this chapter.

The principle of sustainable development and the need for improved integration is not a new debate. In its Global Report on Human Settlements (Planning and Design for Sustainable Urban Mobility) [24], the UN Habitat identified the urban mobility challenge and, more specifically, the sustainability challenge to pivot on the integration of land use and transportation planning; social dimensions and reality; environmental dimensions; economic dimensions, as well as institutional and governmental roles and responsibilities. These principles are directly and indirectly applied in various sources of UN Habitat 2: International Guidelines on Urban and Territorial Planning [25].

From the Africa perspective consisting of developing countries, the State of the African Cities (UN Habitat, 2014) [26] in its '*Re-imagining sustainable urban transitions*' identified integration of factors such as population, urbanization, urban development, urban planning and resource management, urban culture, and green urban development as the main agents for change and development. This implies the involvement of various stakeholders, professions and spheres of government to manage and guide the alignment and integration theme dealt with in this chapter. It is also closely aligned with the focus for UN Habitat 3 [27] on '*The future we want: The city we need*' to be held in 2016.

Many role players within the planning domain have realised that there is a delicate balance between environmental, human, economic, institutional, spatial planning and transportation planning to deliver on sustainable development and to promote alignment and integration. This balance is, however, being impacted upon in a negative sense, due to the fact that this relationship is not addressed and reflected on in the application of planning instruments, tools and implementation by all spheres of government and professions involved. Todes et al. [11] and Berrisford and Kihato [8] provide the reasons for this misalignment. It results in impacts such as ecological degradation, widening of development inequality, segregation and compartmentalization of planning, and non-delivery. It also leads to spatial inefficiency and conflicts in planning and development priorities as being experienced in South Africa.

2 POLICY AND LEGISLATIVE FRAMEWORK GUIDING ALIGNMENT AND INTEGRATION

Tables 1 and 2 give an overview of the policy and legislative framework guiding alignment and integration in terms of policy and legislative framework as applicable to South Africa. The supporting legal principles underpinning the frameworks are included in Scheepers [28] and Van Wyk [29]. The Berrisford and Kihato's [8] paper is clear on the state of considerable flux that exists in the policy and legislative provisions since democratization in 1994. Transformation from the apartheid system to the democratic system represented many challenges to accommodate change in an orderly and legal fashion.

The complexities in alignment between spheres of government, professions and stakeholders can be deduced from an analysis of the existing policy and legislative framework. This diversity and complexities promote and enhance misalignment, lack of integration, subjective application of planning principles and planning tools, and non-delivery in terms of roles and functions. In order to understand this, the transformation of spatial planning, transportation planning and environmental management within spatial systems should be noted.

2.1 Influence of spatial planning transformation

With democratisation in South Africa in 1994, the democratic government inherited a segregated and fragmented spatial system guided by an evenly ineffective policy and legislative framework. The transformation of this framework (notwithstanding several authors referring to the spatial fragmentation consequences of its application by previous political dispensations) commenced with the Draft Green Paper on Development and Planning prepared by the National Development and Planning Commission in 1999 [29]. This was followed by the White Paper on Planning and Land Use Management (RSA, 2001) [5].

In context to the guiding principles for alignment and integration included in these documents, it was preceded by the Development Facilitation Act, Act 67 of 1995 [30], as a vehicle to accommodate spatial change and transformation within all spatial systems. Todes et al. [11] conclude that planning in the post-apartheid period focused on facilitation, with an emphasis on reconstruction and development. Planning instruments such as the IDP process (strategic planning) and the introduction of Spatial Development Frameworks (SDFs) as well as other sectoral plans were introduced.

These processes, although it was a step forward, did not resolve the need to transform the legislative realty guiding spatial planning and development in terms of the new democratic needs and expectations. It created new strategic processes with a general requirement for alignment and integration. More emphasis on the need for alignment and integration, is provided for by SPLUMA [18] and its regulations [19].

Researchers such as Dewar et al. [31], Tomlinson [32], Swilling [33], Mabin and Smit [34], PlanAct [35], Tilman and Wall [36] and Harrison et al. [37] have documented the spatial implications of the historical spatial planning and development practices in South Africa.

Since its inception in 2010, the National Planning Commission (NPC) contributed to documenting the effects of the spatial legacy by the formulation of the Diagnostic Overview [38] and strategic (intervention) proposals as contained in the National Development Plan (NDP) [21]. Oranje and Merrifield [39] reported extensively on national spatial development planning in South Africa (1930–2010), while Drewes and Van Aswegen [40] published an overview of the historic process of national planning in South Africa from a temporal perspective. Lessons learnt from these will assist in applying SPLUMA (2013) through promotion of alignment and integration in context to the theme of this chapter.

Spatial development in South Africa can be subdivided into four distinct spatial development phases:

- Phase 1: Urban and rural formation phase (1652 to 1948).
- Phase 2: Urban and rural separation (fragmentation) phase (1948 to 1994).
- Phase 3: Urban re-integration phase and rural development (post-1994 to 2010).
- Phase 4: Spatial system development, reconstruction, integration and consolidation phase (post-2010 to the present).

Of significance in the transformation process related to spatial planning and development was the promulgation of the first new set of democratic laws that serve as a pivotal point to restructure the characteristics of spatial planning in South Africa and the need for alignment and integration. The spatial planning scene changed significantly with the promulgation of the Constitution [1], Municipal Structures Act [41] and the Municipal Systems Act [42]. This resulted in the formulation of Integrated Development Plans (IDPs) (previously Land Development Objectives (LDOs)) for all institutional entities within the municipal spheres of government. The transition and transformation process culminated in the establishment of the National Planning Commission (NPC) in 2011, the National Development Plan (NDP) in 2012 [21], and eventually the approval of SPLUMA (2013) [18] and SPULMA Regulations (2015) [19].

2.2 Transportation planning transformation to enhance alignment and integration

Transportation planning in South Africa became a statutory planning activity with the enactment of the Urban Transportation Act 78 of 1977 [43]. Transportation planning, development and management are guided by the core policies and legislative framework. The

National Land Transport Transition Act (NLTTA) [44] was used until 2009 when the National Land Transport Act [45] and its Regulations [46], [47] were promulgated.

In this context, various policy documents guide the transport planning, transformation and regulating processes: White Paper on National Transport Policy [48]; Moving South Africa (Vision 2020) [49]; Rural Transport Strategy for South Africa [50]; the National Transport Master Plan 2050 (NATMAP 2050) [20], [49] and the Draft National Land Transport Framework (2014) [23]. In 2014/2015, the National Department of Transport (NDoT) commenced with the revision and adaption of NATMAP 2050 by the formulation of a Synopsis Report [51] to address specific components related to alignment, integration and transportation-related system and network issues and components. It includes addressing the interface between the NDP (2012) [21], SPLUMA [18] and NATMAP 2050 [23], [51] and related areas of concern.

2.3 Environmental management in enhancing alignment and integration

South Africa was slow to develop and institute formal procedures for environmental assessment. It was only with the enactment of the Environment Conservation Act [52] that provision was made to formulate environmental policy to guide decision-making and to prepare environmental impact reports as pointed out by Sowman et al. [53]. The publication of the document entitled *Integrated Environmental Management (IEM) in South Africa* (Council for the Environment, 1989) marked the introduction of the concept of environmental management in South Africa. The term IEM was chosen to indicate a general approach that integrates environmental considerations across all stages of the planning and development cycle and would be applicable to policies, programs, plans and projects [54].

The publication of the National Environmental Management Act (NEMA) [55] introduced a new approach to, and role of environmental considerations in development. NEMA provides the framework for co-operative environmental governance in South Africa and promotes the application of environmental assessment and management tools to ensure integrated environmental management of activities. The intention of NEMA was formalized with the publication of the EIA Regulations (2006). On 3 August 2010, the revised NEMA EIA Regulations [56] were promulgated and includes changes to the listing of activities that impact on the use of land as provided for in SPLUMA [18].

2.4 Need for improved alignment and integration

In assessing the role of alignment and integration within the context of this chapter, Tables 1 and 2 were compiled. The tables summarise core policies and legislative framework as to assess alignment and integration between spatial planning, transportation planning, and environmental management. Most of the policy and legislative framework provides for the promotion of the principles underlying alignment and integration of functions and processes between all spheres of government although in practice planning, implementation and operative processes are being dealt with in isolation. The lack of delivery within all spatial systems is indicative of failing alignment and integration processes notwithstanding a diverse and detailed policy and legislative framework.

The promulgation of SPLUMA (2013) [18], necessitates that the causes and effects of historic weak alignment on implementation and delivery be assessed. No formal surveys will be carried out, as this chapter focuses on assisting and informing basic knowledge on improving alignment, integration, strategic planning and the application of modelling techniques for improved decision making and delivery.

Table 1: Core policies guiding the interface between spatial planning, transportation planning, and environmental management. *(Source: author's own compilation, 2015.)*

Spatial planning	Transportation planning	Environmental management
Reconstruction and Development Plan (1994) Growth, Economic and Redistribution Strategy (1996) White Paper on South African Land Policy (1997) Urban Development Framework (1997) Rural Development Framework (1997) White Paper on Water and Sanitation (1997) White Paper on Local Government (1998) Green Paper on Development and Planning (1999) National Integrated Rural Development Strategy (2000) King Report II on Corporate Governance for South Africa (2002) White Paper on Spatial Planning and Land Use Management (2001) National Spatial Development Perspective (2003) Mining Charter (2003) Construction Charter (2005) ASGISA (2006) National Spatial Development Perspective (2006)	White Paper on National Transport Policy (1996) Moving South Africa (1996) Rural Transport Strategy for South Africa (2003) Draft minimum requirements for the preparation of integrated transport plans (ITP) (2007) NDOT: Public Transport Strategy (2007) NDOT: Public Transport Action Plan (2007–2010) National Land Transport Strategic Framework (2006–2011) (2002) (Draft) NDOT: Road Infrastructure Strategic Framework for South Africa (2006) NDOT: Rural Transport Strategy for South Africa (2007) NDOT: Implementation Strategy to Guide the Provision of Accessible Transport in South Africa (2009) NDOT: Final Draft National Scholar Transport Policy (2009) NDOT: Transport Action Plan (2010) NDOT: National Transport Master Plan 2050 (NATMAP) (2010)	Global Biodiversity Strategy: Guidelines for action to save, study and use earth's biotic wealth sustainably and equitably (published by the WRI, IUCN and UNEP in 1992) Balancing the Scales: Guidelines for increasing Biodiversity's Chances through Bioregional Management (published by the World Resources Institute in 1996). Minimum requirements for the Classification, Handling and Disposal of Hazardous Waste (Second Edition) (1998) (DWAF) Integrated Environmental Management Guidelines Series (1992) DEAT: An Environmental Policy for South Africa (Green Paper) (1996) White Paper on Integrated Pollution and Waste Management for South Africa (2000) Strategic Environmental Assessment in South Africa (2000) DEAT (2002) *Screening, Information Series 1,* Department of Environmental Affairs and Tourism (DEAT), Pretoria. DEAT (2002) *Scoping, Integrated Environmental Management,* Information Series 2, Department of Environmental Affairs and Tourism (DEAT), Pretoria.

Table 1: Continued.

Spatial planning	Transportation planning	Environmental management
Housing Atlas (2006) Sustainable Human Settlement Planning: Resource Book (2008) (NDoH) Area-based Planning. Department of Rural Development and Land Reform (2008/2009) NPC: A Guide to the National Planning Framework (2009) NPC: Green Paper. National Strategic Planning (2009) Comprehensive Rural Development Programme. Department of Rural Development and Land Reform (2009) National Planning Commission Diagnostic Overview (2011)* National Development Plan (2012)* Draft Integrated Urban Development Framework (2015)	NDOT: Road Freight Strategy for South Africa (2011) NDOT: Non-Motorized Transport (NMT) Policy (2012) NDOT: Department of Transport Strategic Plan (2012-2014) TRANSNET: Long-Term Planning Framework (2012) PRASA: PRASA National Strategic Plan (2012) NDOT: Draft National Land Transport Strategic Framework (NLSF) (2014) NDOT: NATMAP Synopsis Report (Draft) 2015*	DEAT (2002) *Specialist Studies, Information Series 4,* Department of Environmental Affairs and Tourism (DEAT), Pretoria. DEAT (2002) *Impact Significance,* Integrated Environmental Management, Information Series 5, Department of Environmental Affairs and Tourism (DEAT), Pretoria. DEAT (2004) *Overview of Integrated Environmental Management,* Integrated Environmental Management, Information Series 0, Department of Environmental Affairs and Tourism (DEAT), Pretoria. DEAT (2004) *Criteria for determining Alternatives in EIA*, Integrated Environmental Management, Information Series 11, Department of Environmental Affairs and Tourism (DEAT), Pretoria. DEAT (2004) *Environmental Impact Reporting,* Integrated Environmental Management, Information Series 15, Department of Environmental Affairs and Tourism (DEAT), Pretoria. Strengthening Sustainability in the Integrated Development Planning Process (2001) State of the Environment Reporting: Draft Guidelines for Local Municipalities (2005) DEAT Information Series (2004-2009) National Framework for Sustainable Development (2008)

*Cross-cutting policies.

Table 2: Core legislative framework guiding interface.** *(Source: author's own compilation, 2015.)*

Spatial planning	Transportation planning	Environmental management
National Building Regulations and Building Standards Act 103 of 1977 Town Planning and Township Ordinance, Ordinance 15 of 1986 Land Use Ordinance (Cape of Good Hope), Ordinance 15 of 1985 Removal of Restrictions Act 84 of 1967 The Less Formal Township Establishment Act 113 of 1991 The Physical Planning Act 88 of 1967 (Sections 6, 8 and 11) Development Facilitation Act 67 of 1995 (DFA) Constitution of the Republic of South Africa 108 of 1996 Bill of Human Rights 1996 Physical Planning Act 88 of 1967 Municipal Structures Act 117 of 1998 Restitution of Land Rights Act 22 of 1993 Interim Protection of Informal Rights Act 76 of 1995 Prevention of Illegal Eviction from Unlawful Occupation of Land Act 19 of 1998 Reconstruction and Development Programme Act 79 of 1998 Municipal Systems Act 32 of 2000 Development Facilitation Act 67 of 1995 (DFA) Physical Planning Act 125 of 1991 Less Formal Township Establishment Act 113 of 1991 (LEFTEA) Subdivision of Agricultural Land Act 70 of 1970 (SALA) Removal of Restrictions Act 84 of 1967 Community Land Reform Act 28 of 1996 (CLARA) Housing Act 107 of 1997 National Land Use Management Bill (Draft 2008) Local Government: Municipal Integrated Development Planning Regulations, 2001. Spatial Planning and Land Use Management Act (SPLUMA) 16 of 2013* R. 239 SPLUMA Regulations (23 March 2015)*	Advertising on Roads and Ribbon Development Act 21 of 1940 Fencing Act 31 of 1963 National Land Transport Transition Act 22 of 2000* Urban Transport Act 78 of 1977 National Transport Interim Arrangements Act 45 of 1998 Transport Appeal Tribunal Act 39 of 1998 Cross Border Road Transport Act 4 of 1998 Road Traffic Act 29 of 1989 National Road Traffic Act 93 of 1996 The South African National Roads Agency Limited and National Roads Act 7 of 1998 National Land Transport Act 5 of 2009 and Regulations (R.1208, 2009) R. 877 National Land Transport Act 5 of 2009: National Land Transport Regulations on Contracting for Public Transport Services	Health Act 63 of 1977 Water Act 54 of 1956 National Water Act 36 of 1991 Water Services Act 108 of 1997 National Environmental Management Act 107 of 1998 (NEMA) National Environmental Management: Air Quality Act 39 of 2004 National Environmental Management: Waste Act 59 of 2009 National Environmental Management: Biodiversity Act 10 of 2004 (NEMBA) National Environmental Management: Protected Areas Act 57 of 2003 (NEMPAA) National Heritage Resources Act 25 of 1999 (NHRA) Mineral and Petroleum Resources Development Act 28 of 2002 (MPRDA) World Heritage Convention Act 49 of 1999 Biodiversity Act 10 of 2004 R.543: National Environmental Management Act 107 of 1998): Environmental Impact Assessment Regulations, 2010 (33306) R.544: Listing Notice 1: List of activities and competent authorities identified in terms of sections 24(2) and 24D (33306) R.545: Listing Notice 2: List of activities and competent authorities identified in terms of sections 24(2) and 24D (33306) R.546: Listing Notice 3: List of activities and competent authorities identified in terms of sections 24(2) and 24D (33306) R.547: Environmental Management Framework Regulations, 2010

*Cross-cutting legislation.

**Various by-laws exist within municipalities.

The transformation reality, as summarized above, is indicative of the complexities and challenges in the alignment and integration of the policy and legislative frameworks. This is a well-known area of debate among professionals. Practitioners in these disciplines within the public and private sectors have interpreted the interface for professional and practical reasons differently and alternatively. The evolution over the past decade of policy and legislative frameworks did not necessarily promote a common understanding of the interface between processes.

From an assessment of the content of the policy and legislative framework (Tables 1 and 2), it can be concluded that there is restricted provision for the formal alignment and integration of the interface between spatial planning, transportation planning, and environmental management. References to this are very general and sometimes of a purely philosophical nature. This statement should be interpreted with the objectives, as identified in the National Framework for Sustainable Development in South Africa (NFSD) (2008) [57] that makes provision for enhancing systems for integrated planning and implementation; sustaining ecosystems and using natural resources efficiently; economic development via investing in sustainable infrastructure; creating sustainable human settlements, and responding appropriately to emerging human development, economic and environmental challenges.

The provisions contained in SPLUMA [18] and the SPLUMA Regulations [19] provide, *inter alia,* for the application of specific development principles and norms (Chapter 2); intergovernmental support (Chapter 3); spatial development frameworks (Chapter 4); land-use management (Chapter 5), and land-development management (Chapter 6). These provisions should be interpreted from the alignment and integration debate, as reported on in this chapter. The lessons learned indicate a lack of alignment and integration in terms of the processes that underpin the current policies and legislative framework. Thus far, SPLUMA [18] seems to be a step ahead in the alignment of spatial planning, land use management, and land development.

However, the existing policy and legislative framework (preceding SPLUMA) is classified as complicated, confusing and incomplete. The lack of a comprehensive and overarching guideline document and processes to promote and integrate planning and development as fields of specialization by various foci can clearly be inferred. Much of the alignment within the policy and legislative frameworks is strategic in nature and does not address functional and operational perspectives.

3 A FRAMEWORK TO MEASURE ALIGNMENT AND INTEGRATION

SPLUMA [18] provides for the following objectives that are fundamental to enhance alignment and integration:

- To provide for a uniform, effective and comprehensive system of spatial planning and land-use management.
- To ensure that the system of spatial planning and land-use management promotes the social and economic inclusion.
- To provide development principles, norms and standards.
- To provide for sustainable and efficient use of land.
- To provide for co-operative government and intergovernmental relations within all spheres of government.
- To redress imbalances of the past and to ensure equity in spatial development planning and land-use management systems [18].

Against this background, Table 3 shows the interface between foci and instruments as provided for in spatial planning, transportation planning, and environmental management policies and legislation (Tables 1 and 2).

In assessing the content of Tables 1, 2 and 3, the complexities and need for alignment and integration in planning within all spatial systems are evident. For the purposes of this chapter, it is not only focuses, processes and strategic approaches that need to be aligned and integrated. The effect of such planning and development within the spatial system needs to be assessed as well. Modelling is the appropriate way as to deal with the effect of planning and development in an accountable fashion.

Attaining this goal depends on the sphere of government responsible for the compilation, approval, implementation and monitoring and may include a national department, a provincial department, or a local government. It may also be a concurrent responsibility between different spheres of government. As will be deduced later in this chapter, the availability of data to undertake modelling to quantify and qualify spatial planning proposals and implementations seems to be problematic in developing countries.

4 IMPACT OF THE DIVISION OF FUNCTIONS, ROLES AND COOPERATION BETWEEN SPHERES OF GOVERNMENT

The role of government in South Africa is based on functions and obligations allocated to the different spheres in terms of the legislative framework (Table 2), intergovernmental cooperation arrangements and Constitutional Court rulings [58]. From an assessment of the

Table 3: Interface in focus and instruments as provided for in core spatial planning, transportation planning, and environmental management legislation. *(Source: author's own compilation from SPLUMA [18], NLTTA [46], NLTA [47] and NEMA [55].)*

SPLUMA (2013)	NLTTA (2000)/NTA (2009)	NEMA (1998)
• Development principles, norms and standards • Intergovernmental support • Spatial Development Frameworks (SDFs) • National Spatial Development Framework (NSDF) • Provincial Spatial Development Framework (PSDF) • Regional Spatial Development Framework (RSDF) • Municipal Spatial Development Framework (MSDF) • Land-Use Management (LUM) • Land-Development Management (LDM) • Municipal Land-Use Plans (MLUP) • Statutory Planning (SP)	• General principles for transportation planning • Types of transportation plans • Provisions on transportation planning • National Land Transport Strategic Framework (NLTSF) • Provincial Land Transport Frameworks (PLTF) • Integrated Transport Plans (ITPs) • Freight Transport Plans (FTP) • Transportation plans and changes in land use and public transport infrastructure and services • Rationalization of public transport services (RATPLANS) • Public Transport Plans (PTPs) • Commuter rail plans (CRPs) • Transport Impact Studies (TISs) • Traffic Impact Assessments (TIAs)	• General objectives • Environmental Implementation Plans (EIPs) • Environmental Management Plans (EMPs) • Integrated Environmental Management (IEM) • Environmental Impact Assessments (EIAs) • Environmental Authorizations (EAs) • Strategic Environmental Assessments (SEAs) • Environmental Management Programmes (EMPs) • Monitoring and Performance Assessments (M&PAs) • Mine Closure Plans (MCPs)

policy and legislation framework, it can be concluded that there are general provisions for the alignment and integration within government and its functions. Provisions for this are contained in the Constitution (1996, Chapter 3, Section 41: 40–41) [1]. The complexities and restrictions in place guiding the division of functions between spheres of government are evident. To understand the division of functions, the obligations to, and roles of co-operative governance should be considered. Cooperative government and intergovernmental relations provide for the following arrangement in terms of alignment and integration:

- To provide and enhance effective, transparent, accountable and coherent government within the national spatial system as a whole.
- To respect and recognize the constitutional status, institutions, powers and functions in all spheres of government.
- To exercise powers and functions conferred on it in terms of the Constitution.
- To exercise the powers and its functions in a manner that does not encroach on the geographical, functional or institutional integrity of any government in another sphere.
- To co-operate with each other in mutual trust and good faith (Constitution, 1996, Chapter 3, Section 41) [1].

The division of functions fulfils an important role as far as alignment and integration are concerned. The role of any sphere of government in terms of functions and the level of responsibility may be classified as strategic, functional, or operational, or any combination thereof.

Several authors contributed further insight into this complex arrangement: National Treasury, *Trends in Intergovernmental Finances: 2000/01–2006/07* [59]; Steytler and Fessha, *Defining Provincial and Local Government Powers and Functions* [60]: *The Management of Concurrency* (2005); COGTA, *Division of Functions* (2002) [61] and Van Wyk, *Planning in all its (Dis)Guises: Spheres of Government, Functional Areas and Authority* [62].

The University of the Western Cape (Community Law Centre) [60] provided a valuable contribution in determining the appropriate functions and powers within local government. Van Wyk [62] concludes that it is an ongoing debate regarding the content of the legislative and executive functional areas relating to 'planning' that are enjoyed by each sphere of government. For alignment and integration to be promoted, the mechanism of cooperative governance must be applied and practiced among all spheres of government. Berrisford and Kihato [8] states specifically that there is a need to create effective intergovernmental communication.

In an effort to simplify the understanding of the complexity of the division of functions, Table 4 was compiled, indicating the core functions and activities allocated to the national, provincial and municipal spheres of government. Table 4 is based on the provisions of the Constitution [1] and the Municipal Structures Act to illustrate the competencies, roles and functions between spheres of government as it was originally intended by the legislator.

Van Wyk [29] points out that Chapter 3 of the Constitution should constantly remind every organ of state, inclusive of every functionary in every sphere of government, of the importance of co-operation, because the principles of co-operation reinforce the values underlying open, transparent and responsible government. Van Wyk further concludes that, in light of the varied functional areas such as agriculture, environment, housing and transport that play a role in planning, the challenge remains to ensure that principles of co-operative government feature significantly at all times. Co-operation remains a constitutional obligation.

Table 4: Core functions and activities allocated to national, provincial and municipal spheres of government related to spatial planning, transportation planning, and environmental management.*

Powers, functions and/or activities			
Schedule 4: Functional areas of concurrent national and provincial legislative competence (Part A)			
Function/activity	National	Provincial	Local
Airports other than international and national airports	x	x	
Environment	x	x	
Nature conservation	x	x	
Pollution control	x	x	
Provincial public enterprises	x	x	
Public transport	x	x	
Public works	x	x	
Regional planning and development	x	x	
Road traffic regulation	x	x	
Urban and rural development	x	x	
Vehicle licensing	x	x	
Schedule 4: Functional areas of concurrent national and provincial legislative competence (Part B)			
Air pollution		x	x
Municipal airports		x	x
Municipal planning			x
Municipal public transport		x	x
Municipal public works		x	x
Pontoons, ferries, jetties, piers and harbours		x	x
Schedule 5: Functional areas of exclusive provincial legislative competence (Part A)			
Provincial planning		x	
Provincial roads and traffic		x	
Schedule 5: Functional areas of exclusive provincial legislative competence (Part B)			
Fences and fencing		x	x
Municipal roads			x
Noise pollution		x	
Street trading			x
Street lighting			x
Traffic and parking			x

*1. Functions and activities should be interpreted in terms of concurrent responsibilities, as contained in the applicable legislative framework; 2. Concept of distinctive, interdependent and interrelated across some functions and activities listed in the Constitution (1996) applies; 3. The rulings of the Constitutional Court as far as powers and functions are concerned, should be consulted; 4. Functions may cut across spheres of government.

Berrisford and Kihato [8] state that there is uncertainty as to the roles of the three spheres of government. The negative effects resulted from the DFA (1995) and the White Paper on Spatial Planning and Land-Use Management (RSA, 2001) [5] that are now being addressed through SPULMA (2013). It, thus addresses the incomplete work of local government reform in South Africa.

Table 5 shows examples of plans formulated by the different spheres of government in terms of powers, functions, duties and activities dealt with earlier and in terms of the focus of this chapter.

Table 5: Planning instruments formulated and managed by different spheres of government.*

Description/type	National**	Provincial	Local***
Transportation planning			
NATMAP 2050	x		
NATMAP 2050 Synopsis Report (2015)	x		
NLTF (2015) (Draft)	x		
Public Transport Plans (PTPs)		x	x
Commuter Rail Plans (CRPs)	x	x	
Rationalization of Public Transportation Services (RAT Plans)		x	x
Freight Transport Plans (FTPs)		x	x
Transportation Sector Plans (TSPs)		x	x
Transportation Infrastructure Plans (TIPs)	x	x	x
Traffic Impact Assessments (TIAs)			x
Transportation Impact Studies (TIS)		x	x
Provincial Land Transport Frameworks (PLTFs)	x	x	
National Land Transport Strategic Framework (NLSTF) (2014)	x	x	
Integrated Transport Plans (ITPs)		x	x
Spatial planning			
National Spatial Development Plan (NDP)	x		
National Spatial Development Framework (NSDF)	x		
Provincial Spatial Development Perspective (PSDP)	x	x	
Regional Spatial Development Framework (RSDF)		x	x
Municipal Spatial Development Frameworks (MSDFs)			x
Land-Use Management Systems (LUMS)			x
Land Development Management Plans (LDMPs)			x
Rural Development Plans (Strategies) (RDPs)		x	x
Municipal Regeneration Plans (Urban Renewal, etc.) (MRPs)			x
Precinct Plans (PPs)			x
Site Development Plans (S.Dev.Ps)			x
Integrated Development Plans (IDPs)		x	x
Integrated Urban Development Framework (IUDF)		x	x
Statutory Planning Processes (SPPs)			x
Spatial Development Plans (SDPs)		x	x
Environmental management			
Strategic Environmental Assessments (SEAs)		x	x
Integrated Environmental Management Plans (IEMPs)		x	
Environmental Management Plans (EMPs)		x	x
Environmental Impact Assessments (EIAs)		x	x
Environmental Authorizations (EAs)		x	
Mine Closure Plans (MCPs)	x	x	
Environmental Implementation Plans (EIPs)		x	x
Environmental Management Programme (EMProg.)		x	
Monitoring and Performance Assessment Plans (MPAPs)		x	x

*Types of plans are generic only. Terminology and content are determined by policies, guidelines and standards.
**May include agencies' concurrent functions with provinces.
***Municipalities are the only sphere of government constitutionally to make and adopt detailed by-laws on land-use management and its other function, as included in the Constitution (1996) and the Municipal Structures Act (1998).

One can infer the need for alignment and integration from Tables 4 and 5. The context should, however, be interpreted in line with Figs 1 and 2. The system complexities, dynamics and need for integration to optimise development within spatial systems are evident.

Fig. 1 illustrates the vertical and horizontal alignment and integration among the three spheres of government. It shows the organization of the different spheres of government, integration and formulation of planning instruments, structures, institutions and agencies involved. It depicts the basic need to promote cooperative governance through both process and management practices within all spheres of government. Various professions as well as the integration of planning instruments (plans) fulfil a vital role in this instance.

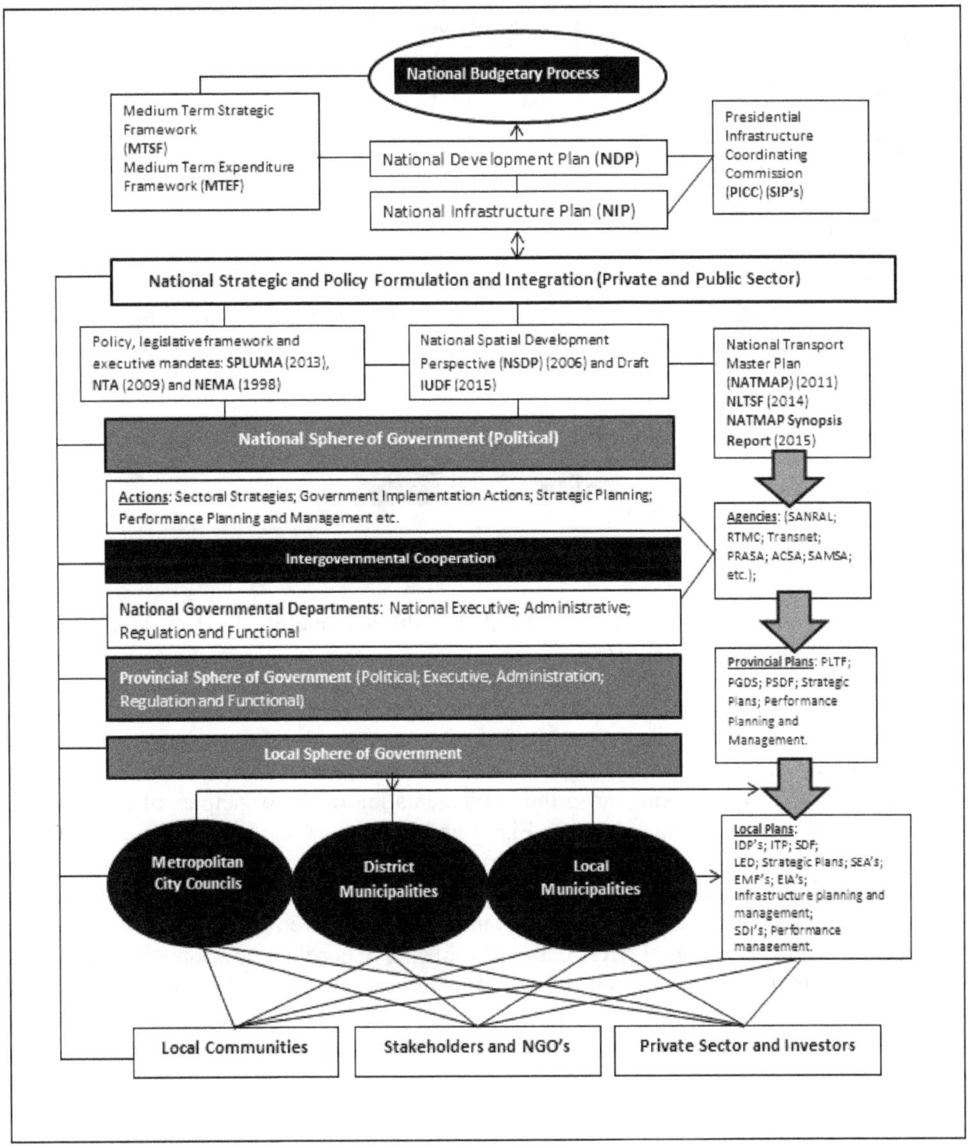

Figure 1: Complexities of alignment and integration between different spheres of government, agencies, co-operation structures and planning instruments. *(Source: author's own compilation, 2015.)*

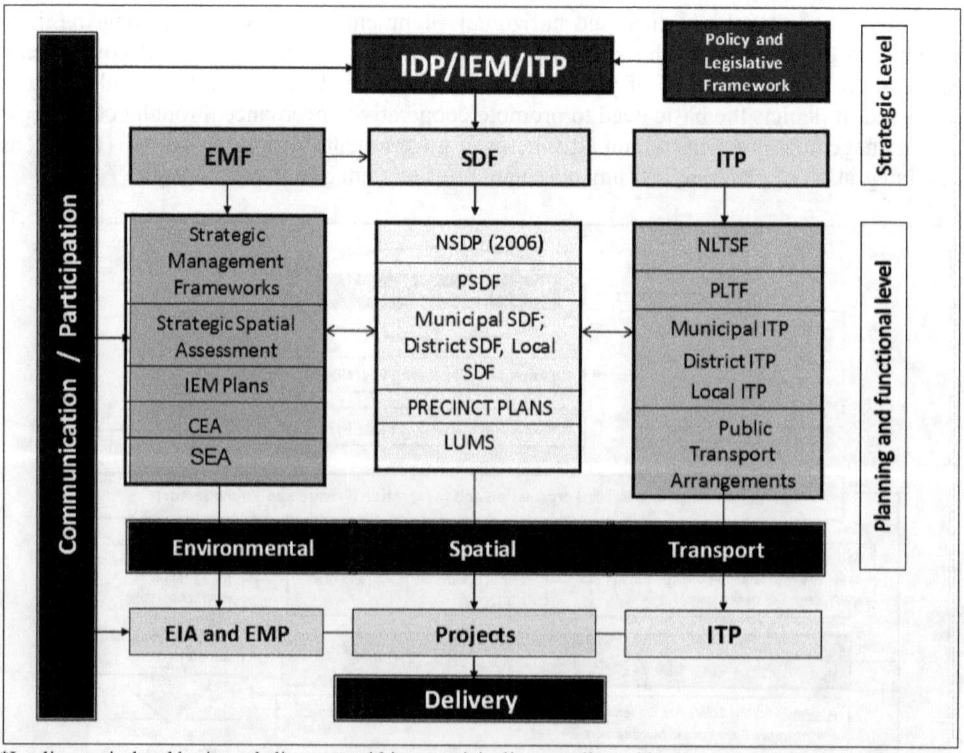

*Implies vertical and horizontal alignment within a municipality as well as with adjacent municipalities. Alignment and integration as provided for in SPLUMA (2013).

Figure 2: Interfaces between IDP, IEM and ITP within municipalities.* *(Source: author's own compilation, 2015.)*

Malan [63] states that the system of intergovernmental relations and co-operative government in South Africa is rapidly evolving. This is due to the statutory commitment of the various spheres of government to the implementation of the principles of co-operative and intergovernmental relations (refer to Figs 1 and 2).

In terms of section 41: 40–41 of the Constitution (1996) [1] (refer to Fig. 1), government is constituted as national, provincial and local spheres of government: distinctive, interdependent and interrelated. Provision is made for intergovernmental structures such as the Intergovernmental Forum (IGF), the President's Coordinating Council (PCC) and Intergovernmental Relations Committees of Ministers and Provincial Councils (MINMEC); Organized Local Government (SALGA) and Forum for South African Directors-General (FOSAD) (COGTA, 2002) to promote alignment and integration between all spheres of government.

5 THE DOMAIN OF SPATIAL PLANNING FROM AN ALIGNMENT AND INTEGRATION PERSPECTIVE: ROLE AND IMPACT OF VARIOUS DISCIPLINES

Generally, planning (including urban and regional planning, transportation planning, and environmental management) entails the consideration of what can and should happen where, in the spatial systems. It includes the foci and interaction of different policies and practices (tools and instruments) across regional space, and sets the role of spaces, places and interaction between professions in a wider context. It goes well beyond 'traditional planning' (inclusive of authorizations) and includes strategic, functional, operational planning and monitoring of all planning and activities. The alignment and integration is related to vertical, horizontal, concentric and three dimensional implications of spatial, transport and environmental on each other. This provides for an ongoing assessment of trans-disciplinary impacts and intervention as to promote sustainable planning and development.

5.1 Role of education and training in professions and capacity

Fig. 3 shows the interfaces between fields of education and training for professions involved in the theme of this chapter. The core relationships between subject themes and disciplines are depicted in terms of their orientation from an urban and regional planning perspective.

The relationship between education, training and professional development is fundamental for the way in which alignment and integration is being dealt with in practice and in the workplace. In terms of SPLUMA (2013), planning includes various disciplines, as illustrated in Fig. 3. In some instances, it may imply duplication within certain spheres of government that requires centralization. Co-operation between planning-related and other disciplines is thus a focal consideration in promoting alignment and integration. Todes et al. [11] state that the need for integration may be addressed through:

- Centralizing disciplines such as planning and environmental management.
- Mainstreaming certain concerns and concepts.
- Institutional integration.
- Coordination through improved vertical and horizontal linkages.

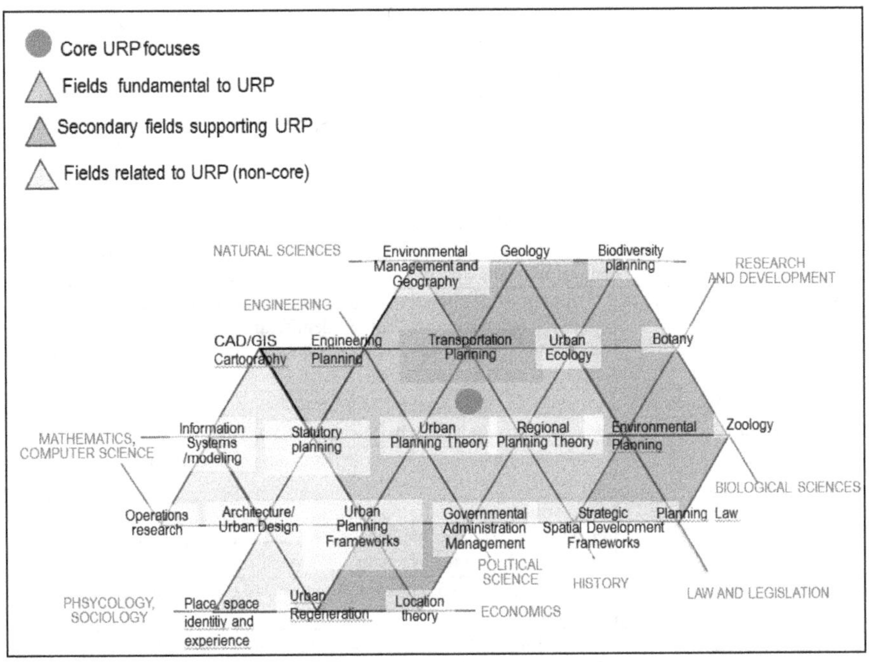

Figure 3: Interfaces in education and training of professions. *(Source: Schoeman, 2010 [12].)*

Table 6 shows a classification of the core foci for planning activities based on international and national norms and principles based on the focuses as dealt with in this chapter. The foci are grouped as follows: spatial planning; urban planning; policy and strategy formulation;

Table 6: Core planning foci based on international and national norms and principles. *(Source: author's own compilation, 2015.)*

Core professional planning focus	Domain as interpreted internationally and nationally
Spatial planning	Planning systems; practices in regional space; role of places; strategic frameworks; forward planning; scale of regional planning; development in physical environment; spatial plan formulation; impact of migration; regional spatial planning needs; regional corridor and nodal development.
Urban planning	Role of places; anticipating development; scale of urban planning; surface and beneath surface development; urban development; urban regeneration and development; urban design; site planning; urban spatial planning needs; neighbourhood development; urban corridor and activity node development; urban renewal.
Policy and strategy formulation	Interaction of policies; policy interventions; multi-perspective approaches; disaster preparedness plans; input in drafting of policy and legislation.
Land-use management	Land-use planning; land-use management and control; regulating development; control of land use; management of change in land use; legal issues related to land use and building codes; legal issues related to environmental regulations.
Built environment	Style of buildings; design of public spaces; conservation of historic buildings; development of public spaces and places; location, design and layout of buildings.
Land availability	Land reservation; identification of land for development.
Transportation planning	Innovative forms of transport; accessibility between places of residence, work and amenities; traffic congestion management; air pollution management; transport and land-use models; transportation frameworks.
Environmental management	Relationship between built and environment; negative impacts on natural environment; natural impacts on communities; protection of natural environments; standard of environmental quality; environmental sustainability; landscape development; legal issues related to environmental management.
Socio-economic and spatial development	Social and economic status quo and forecasting; community regeneration; regional and economic development; rural enterprise; sectoral policies; planning research; technical analysis; smart growth strategies; economic development plans; development of resources; socio-economic profiles.
Facilitation and communication	Compromise formulation; lead public consultation processes; education, training and capacity-building; identification of community needs; community goals and vision compilation; development consultation; public address, meeting and facilitation.
Human settlement development	Housing development; housing strategies.
Rural development	Community development; area-based planning.
Feasibility studies	Appreciation of spatial complexities; deeper underlying causes; integrated analysis.
Implementation	Infrastructure needs; infrastructure programming; general management; needs prioritization; implementation and enforcement strategies; determination of infrastructure and amenities capacity.
Project management	Management of programs for planning and implementation; quality management.
Infrastructure planning	Supporting the planning and development of engineering infrastructure based on sustainable planning policies, practices and needs.
Management and analysis support systems	GIS applications and techniques; modelling; systems analysis.

land-use management; involvement in the built environment; land availability; transportation planning; environmental management; impact assessments; authorizations; socio-economic and spatial development; facilitation and communication; human settlement development; rural development; feasibility studies; implementation; project management, as well as management and analysis based on the application of decision support systems.

The problematic nature of alignment and integration between specific disciplines originates in the education and training of professionals functioning in these domains. The curricula within these programs do not allow for an understanding of the realities being shared among professions. This creates a professional backlog between professions to reason and debate across professional boundaries from an alignment and integration background. This complicates co-operation, plan formulation, implementation and monitoring.

This implies the need for the revision of the competencies and standards applied in curricula for education and training of professions functioning in such disciplines. This is not only a challenge for higher education institutions, but also implies the role and responsibilities of Professional Councils governing such disciplines. The three foci dealt with in this chapter are being controlled by the following professional councils: South African Council for Planners (SACPLAN) (Planning Professions Act 36 of 2002; Engineering Council of South Africa (ECSA) (Engineering Professions Act 46 of 2000), and the South African Council for Natural Scientific Professions (SACNASP) (Natural Scientific Professions Act 27 of 2003). Todes et al. [11] conclude that "the language used in each profession is not necessarily readily understood by the other". This also applies to the theme of this section.

Capacity within disciplines is also a problem for the majority of professions in South Africa. Disciplines not only consist of professionals, but are also supported by an array of supporting staff with other related and sometimes different qualifications that need to be trained, capacitated and applied in disciplines such as spatial planning, transportation planning, and environmental management. Enhancing alignment and integration thus depends on the dynamic reality inclusive of internalities and externalities being influenced by various professions. To this reality the impact of political will and preferences (notwithstanding the policy and legislative frameworks) complicates the attainment of alignment and integration.

The combination or integration of functions or disciplines alone will not resolve such challenges. The solution lies somewhere between a combination of training and education, extension of competencies and standards for professional registration and provision of adequate capacity in terms of specialization and supporting staff. These resources should be applied in the context of the functions, roles and obligation for government to co-operate, as discussed earlier. Capacity building for politicians within all spheres of government on the need for alignment and integration based on empowerment on the principles guiding the related professions based on the policy and legislative framework need be implemented to ensure alignment and integration as discussed in this chapter.

6 SPLUMA (2013) AS AN INSTRUMENT TO PROMOTE ALIGNMENT AND INTEGRATION

The context of addressing the challenge of alignment and integration by SPLUMA [18] is illustrated in Fig. 4, showing the reality to meet the overarching goals and objectives from a sustainability perspective set out in the objectives contained in Chapter 1 of SPLUMA and, specifically, the development principles and norms and standards (Chapter 2).

Table 7 shows the outcome of an alignment and integration analysis based on the provisions of SPLUMA (2013).

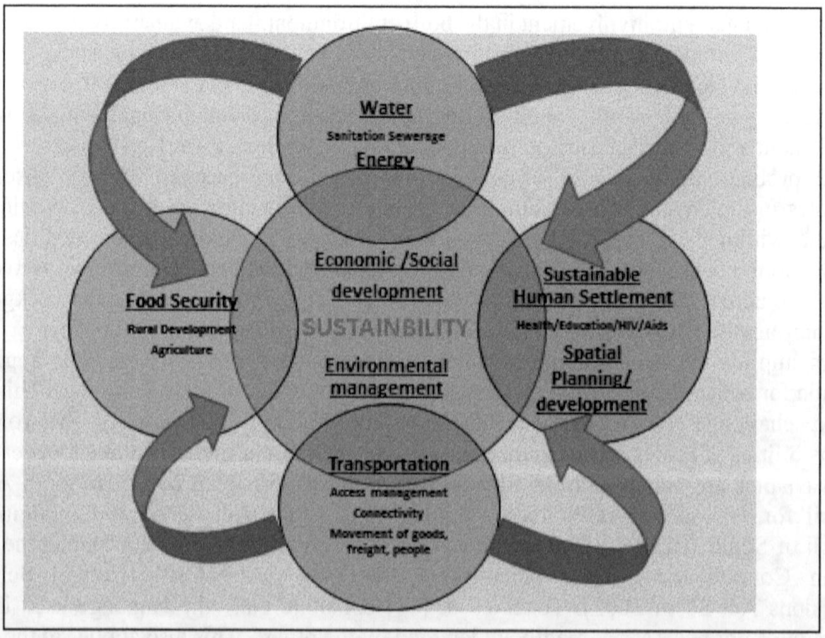

Figure 4: Enhancement of sustainability through alignment and integration by SPLUMA (2013). *(Source: author's own compilation, 2015.)*

The South African Cities Network [64] summarized SPLUMA as a tool for spatial transformation in a very effective and focused manner by explaining the background, and spatial transformation and by identifying potential spatial transformation levers.

Berrisford and De Visser [65] state that SPLUMA is classified as framework legislation rather than as comprehensive revision of the status quo for land-use management. Berrisford and De Visser (2015) further point out that SPLUMA returns to the planning in the pre-1990s thinking, inclusive of the philosophy '*planning everything*'. It relies heavily on SDFs in all spheres of government. It also includes *'wall to wall'* land-use management schemes. It is pointed out that the negative effects resulting from the contents of the DFA (1995) [30], White Paper on Spatial Planning and Land-Use Management [5] and Draft Land Use Management Bill [66] are now being addressed through SPLUMA (2013) and its Regulations (2015). It thus addresses the incomplete work of local government planning reform (Berrisford and Kihato [8]).

SPLUMA (2013) recognizes provincial competencies for law-making (Schedule 1 of the Constitution) and specifies the roles (inclusive of intergovernmental support) of national, provincial and municipalities (Chapters 2 to 4). Chapters 5 and 6 deal with land-use management and land-development provisions and arrangements. Section 8 (Norms and Standards) may be determined by the National Minister to allow for consistency across the country. This will promote alignment and integration.

SPLUMA (2013) further operates parallel to other laws. It repeals existing national laws (see Table 7). SPLUMA adds to what the Municipal Systems Act (2000) provides for in terms of the IDPs and SDFs, and it should be noted that SPLUMA does not contradict the MSA (2000) or any other act.

Table 7: Assessment of roles, alignment, integration and involvement of spheres of government in the implementation of SPLUMA (2013). *(Source: author's own compilation, 2015.)*

Alignment components, functions and obligations	Sphere of government				
	National	**Provincial**	**Metros**	**DMs**	**LMs**
Chapter 1: Introductory provisions					
• Application of the Act	Align	Coordinate	Integrate	Integrate	Integrate
• Objectives of the Act	Align	Integrate	Implement	Integrate	Integrate
• Spatial planning system	Align	Implement	Implement	Implement	Implement
• Categories of spatial planning	Align	Coordinate /Formulate	Formulate	Coordinate/ Formulate	Formulate
Chapter 2: Development principles and norms and standards					
• Application of development principles	Monitor	Integrate	Implement	Implement	Implement
• Development principles	Enacted	Integrate/Apply	Apply	Apply	Apply
• Norms and standards	Enacted	Integrate	Apply	Apply	Apply
Chapter 3: Intergovernmental support					
• National support and monitoring	Apply	Coordinate	Align	Align	Implement
• Provincial support and monitoring	Engage	Implement	Align	Engage	Engage
• Municipal differentiation	Apply	Engage	Align	Apply	Apply
Chapter 4: Spatial development frameworks					
• Preparation of spatial development frameworks	Enacted	Apply	Apply	Apply	Apply
• Preparation and content of national spatial development frameworks	Apply	Engage	Engage	Engage	Engage
• Preparation and content of provincial spatial development framework	Coordinate	Implement	Input	Input	Input
• Preparation and content of regional spatial development framework	Enacted	Coordinate	Implement	Implement	Input
• Preparation and content of municipal spatial development framework	Enacted	Coordinate	Implement	Implement	Implement
• Status of spatial development frameworks	Enacted	Monitor	Enforce	Enforce	Enforce

Table 7: Continued.

Alignment components, functions and obligations	Sphere of government				
	National	Provincial	Metros	DMs	LMs
Chapter 5: Land-use management					
• Role of executive authority	Enacted	Coordinate	Implement	Implement	Implement
• Land-use scheme	Enacted	Coordinate	Implement	Implement	Implement
• Purpose and content of land-use scheme	Enacted	Enforce	Implement	Implement	Implement
• Legal effect of land-use scheme	Enacted	Coordinate	Enforce	Enforce	Enforce
• Review and monitoring of land-use scheme	Enacted	Coordinate	Enforce	Enforce	Enforce
• Amendment and monitoring of land-use scheme and rezoning					
• Consultation with other land development authorities	Enacted	Coordinate	Management	Management	Management
• Alignment of authorizations	Enacted	Coordinate	Engage	Engage	Engage
• Record of amendments to land-use scheme	Enacted	Coordinate	Engage	Engage	Engage
• Enforcement of land-use scheme	Enacted	Monitor	Compile	Compile	Compile
	Enacted	Coordinate	Apply	Engage	Apply
Chapter 6: Land-development management					
• Municipal land-use planning	Enacted	Coordinate	Implement	Implement	Implement
• Municipal cooperation	Enacted	Coordinate	Engage	Engage	Engage
• Establishment of Municipal Planning Tribunals	Enacted	Monitor	Implement	Implement	Implement
• Processes of Municipal Planning Tribunal	Enacted	Coordinate	Implement	Implement	Implement
• Related land-development matters	Enacted	Monitor	Engage	Engage	Engage
Chapter 7: General provisions					
• Commencement of registration of ownership	Enacted	Coordinate	Implement	Implement	Implement
• Regulations	Enacted	Formulate	Apply	Apply	Apply
• Exemptions	Enacted	Coordinate	Determine	Determine	Determine
• Delegation	Enacted	Coordinate	Apply	Apply	Apply
• Non-impediment of function	Enacted	Coordinate	Monitor	Monitor	Monitor
• Offences and penalties	Enacted	Coordinate	Apply	Apply	Apply
• Repeal of laws	Enacted	Coordinate	Apply	Apply	Apply
• Transitional provisions	Enacted	Coordinate	Implement	Implement	Implement
• Short title and commencement	Enacted	Implement	Apply	Apply	Apply

Table 7: Continued.

Alignment components, functions and obligations	Sphere of government				
	National	**Provincial**	**Metros**	**DMs**	**LMs**
Schedule 1: Matters to be addressed in Provincial Legislation					
• Provincial legislation regulation land development, land-use management, township establishment, spatial planning, subdivision of land, consolidation of land removal of restrictions and related matters	Enacted	Implement	Apply	Apply	Apply
Schedule 2: Schedules land-use purposes					
• List of land-use purposes	Enacted	Implement	Implement	Implement	Implement
• Definitions	Enacted	Apply	Apply	Apply	Apply
Schedule 3: Appeal of laws					
• Removal of Restrictions Act (1967)	Enacted	Coordinate	Transition	Transition	Transition
• Physical Planning Act (1967)	Enacted	Coordinate	Transition	Transition	Transition
• Less Formal Township Establishment Act (1991)	Enacted	Coordinate	Transition	Transition	Transition
• Physical Planning Act (1991)	Enacted	Coordinate	Transition	Transition	Transition
• Development Facilitation Act (1995)	Enacted	Coordinate	Transition	Transition	Transition

7 PERSPECTIVES ON STRATEGIC PLANNING AND MODELLING AS MECHANISMS TO GUIDE ALIGNMENT AND INTEGRATION

The line of argument for improved alignment and integration between spatial planning, transportation planning and environmental management within spatial systems (in both developing and developed countries) is based on the need for a new approach and/or mechanism to promote such goals and objectives. The approaches as formulated by Geerlings and Stead [2], Stead et al. [3] and Stead and Meijers [67] focused on policy development that is applicable to the theme being dealt with in this chapter. However, the complexities in terms of policies directing the focuses in South Africa is much more diverse and complicated (see Table 1 and Table 2) and the line of argument developed in this chapter are based on these complexities.

Strategic planning through integrated development planning applied to South Africa in [17] (an example for other developing countries) provides an applicable framework to deal with such complexities in terms of policies and spatial planning responsibilities such as alignment and integration. Table 8 provides an overview of the content included in the integrated development plan for South Africa to promote alignment and integration [68]. From the assessment of the abovementioned content the absence of sector plans such as an Environmental Management Framework (EMF), Integrated Environmental Management (IEM) and Strategic Environmental Management Assessment (SEA) is evident. This table should also be related to the contents as included in Figs 1 and 2.

Alignment and integration in this context within all other spheres of government is evident and needs to be implemented and coordinated amongst all spheres of government (within the context of planning, identification of needs, prioritization, implementation and operationalization).

This interpretation also needs to be assessed against the following approaches supporting strategic planning as a mechanism to promote alignment and integration:

Albrecht et al. [69] and Albrecht [70] state that in the 1990s the strategic approach to the organization of space at different levels of scale became more prevalent (note: the IDP process in the RSA as discussed above was formalized since 2002). He concludes that 'The 'alternative' strategic planning presented in this paper is conceived as a democratic, open, selective, and dynamic process. It produces a vision to frame problems, challenges, and short-term actions within a revised democratic tradition'. Albrecht further points out that the strategic process is underlain by the following factors:

- It involves content and process.
- Statics and dynamics.
- Constraints and aspirations.
- The cognitive and the collective, the planned and the learned.
- The socioeconomic and the political.
- The public and the private.
- The vision and the action.
- The local and the global.
- Legitimacy and a revised democratic tradition.
- Values and facts.
- Selectivity and integration ability.
- Equality and power.
- Long term and short term.

Table 8: Summary of the Integrated Development Plan (IDP) Process implemented in South Africa to address the complexities in the spatial systems from a strategic planning perspective. *(Source: adapted from DPLG, 2008 [66].)*

	PACKAGING OF THE IDP: IDP COMPONENTS *"A plan for all Government"*
A	**Executive Summary:** A concise overview of the municipality, its current situation, key challenges and opportunities, as well as the priority strategies of the municipality to improve its situation over the five-year term of the IDP, indicating the most critical targets to be achieved. This summary should answer the following questions: 1. Description of the institution 2. Challenges and issues to be addressed 3. What opportunities exist 4. An approach for improvement and integration 5. Next five (5) years planning focus 6. Measurement of progress (application of balanced score cards (BSC) and goal achievement matrixes (GAM)) 7. IDP development
B	**Situational Analysis:** A detailed (deep rigorous) status quo analysis of the municipal area, with Inputs from the Community, Community Organizations, Business, Non-State Actors, Sector Departments, SOE's – State Owned Enterprises, and Spatial Analysis. SWOT analysis critical environmental scan. The use of latest acceptable information and data should inform the analysis. This section to be concluded by Identification of Priority Issues in relation to: *Demographic, Economic, Infrastructure, and Environmental issues. This section also to reflect which entity is responsible for what service in terms of powers and functions.*
C	**Development Strategies:** Long Term Growth and Development Goals (Vision, Mission, Strategic Objectives and Strategies structured into 5 plus Key Performs Areas (KPAs) of the Five Year Local Government Strategic Agenda. *(Suggested to be forecasted at 10–15 Years and structured into the 5 KPAs of the Five Year Local Government Strategic Agenda)* The Additional Area would be Spatial Analysis.
D	**High Level Spatial Development Framework (SDF):** Overview of Spatial Development Framework (Text and few Maps)
F	**Implementation Plan: 5-3 Year** Five Year, Three Year Implementation Plan with Committed Human and Financial Resources. Developed to Project level. *(To be Updated annually)*
G	**Annual Operational Plan – Business Plan** One Year Detailed Operational Plan that speaks to SDF. The SBDIP and Annual Plan should become one plan in the applicable Financial Year. *(Reviewed Annually)*
H	**Financial Strategy (Financial Plan)** Multi-Year Budget with a three-year commitment and a strategy for municipal revenue generation, outside grants (funds), over a medium to long term. The Annual Budget should be confirmed and adopted with the IDP. *(Reviewed annually)*

Table 8: Continued.

	PACKAGING OF THE IDP **IDP COMPONENTS** *"A plan for all Government"*			
I	**Organisational Performance Management System:** Key Performance Indicators for each objective and Annual Performance Report of previous year. *(Reviewed Annually)*			
J	**ANNEXURES:** This section to be prefaced by an outline on the status of all applicable Sector Plans. This can be done in a single table.			
	Sector Plan Focus	**Applicable to Sphere of Government**		
		Metropolitan Councils	**District Municipalities**	**Local Municipalities**
J.1	Detailed Spatial Dev. Framework,	Yes	Yes	Yes/No
J.2	Detailed Disaster Management Plan	Yes	Yes	Yes/No
J.1	Land Use Management Framework	Yes	Yes	No
J.2	Land Use Management System	Yes	No	Yes
J.3	Coastal Zone Management Plan	Yes	Yes	Yes/No
J.4	Waste Management Plan	Yes	Yes	Yes
J.5	A. Water Service Development Plan (all WSAs); B. Water Resources Plan; C. Forestry Plan	Yes	Yes	Yes/No
J.6	Integrated Transport Plan	Yes	Yes (Bulk)	Yes
J.7	Housing Plan	Yes	*No*/yes	Yes
J.8	Energy Master Plan (Electricity Master Plan)	Yes	Yes (Bulk)	Yes
J.9	Local Economic Development Plan	Yes	Yes	Yes/no
J.10	Infrastructure Investment Plan Extended Public Work Program (EPWP) Compliant. The Infrastructure Plan is derived from the Long-Term Growth and Development Strategy, and is primarily based on the rigorous spatial analysis. It would therefore be contextualized with the District's Spatial Development Framework, and the Land Use Management System of the Local Municipality. A considered and well-designed Infrastructure Plan will ensure the delivery of services and development of the local economy.	Yes	Yes	Yes
J.11	Area Based Plans (Land Reform)	Yes	Yes	No
J.12	Anti-Corruption Strategy	Yes	Yes	Yes
J13	**What is needed from Sectors** This must take the simple form of a table. It needs to list the sector department and what is needed from that department in terms of assistance, support, information, etc.	Yes	Yes	Yes

Olesen [71] has taken the debate further to the neo-liberalisation of strategic spatial planning. She states that neo-liberalisation of strategic planning has manifested as a result of governance reforms aiming to reduce or abolish spatial planning at national and regional scales. This is in stark contrast to the need for strategic planning as to integrate transportation and spatial planning as to optimize and enhance alignment and integration to promote delivery and sustainability (Fig. 4). This action is also in contradiction to the application of planning instruments as being dealt with in this chapter.

Modelling as mechanism to enhance alignment and integration refers to the endeavour to enhance the interface through representing the reality of spatial planning, transportation planning and environmental management in an abstract way by relating the focuses of the planning instruments (plans) to each other with the view to predict impacts and long term predictions.

Several authors have reported on integrating the environment in urban planning and environment. The most valuable contribution to this chapter relates to the report published by the United Nations Environmental Programme (UNEP) in 2013 containing key principles and approaches for cities in the twenty-first century [72].

Other publications and approaches more specifically deal with the effects of urban form and/or land use on transportation planning and *vice versa* [73]–[75]. Related research themes focused on the development of a framework to assist the integration of environmental, social and economic issues in spatial planning [76]. Research related to the question of spatial planning and environmental assessment [77] supports the theme as discussed in this chapter. However, limited specific research thus far focused on the alignment and interface as discussed in context to the line of argument above.

In support of the theme of this chapter, McKenna Davis (Editor) in 2010 [78] published an in-depth case analysis related to green infrastructure. Of significance to the theme of this chapter is the application of mapping for planning from a sustainability perspective.

The best example of modelling complexities in alignment and integration between spatial, transportation and environmental planning and management in support of the abovementioned principles are contained in the work by Waddell in 2011 [79]. He illustrates fundamentally how land use and transportation planning modelling and its challenges in research and practice be approached. He points out that academic research in integrated land use planning and transport is on the rise. This statement by implication links closely to the need for the integration in context to the focus of this chapter. However, alignment and integration consists of much more than to predict outcomes but to optimize space and places in terms of planning, development and delivery. In this the factors of spatial balance, potential realization, continuity and sustainability are the main considerations.

Modelling is basic to strategic planning and decision making and present practices need to be re-engineered as to support the domains of alignment and integration. This will require that the outcome and implications for all disciplines, plans and development focuses be modelled in terms of alternative scenarios and that the related impacts are appropriately quantified and qualified.

Furthermore, to enhance alignment and integration, fundamental changes in management and planning approaches and practices (inclusive of addressing the issue of contradictions, non-alignment of planning processes and duplication in procedures) need to be identified in system wide contexts with the common objective to promote sustainability on trans-disciplinary levels.

Waddell [79] identifies the following challenges for integrated modelling:

- Transparency.
- Behavioural validity.

- Empirical validity.
- Ease of use or simplicity.
- Computational performance.
- Flexibility.
- Data availability and quality.
- Role of uncertainty.

In his article [16], Waddell applies UrbanSim in modelling urban development for land use, transportation and environmental planning. This is possible in developed countries where the addressing of the abovementioned challenges could be met. In developing countries, only some of the attributes used in UrbanSim restricts successful application to asses alignment and integration within the spatial systems. In Chapter 2 the authors show how to overcome these restrictions in a developing country.

This implies the development of a more simplified macro modelling approach to address the challenge of alignment and integration. The methodologies as discussed above need to be applied in an integrated fashion to address the dynamics such as interest of different sphere of government, institutions, stakeholders, values, policies, legislation, guidelines and planning within all spatial systems in an integrated approach.

In developing such a modelling approach the spatial effects of spatial planning, transport planning and environmental management (plans) will have to be applied, quantified and qualified based on a model that are able to optimize spatial alignment, integration and impacts in support of sustainability (Fig. 4).

8 PERSPECTIVES FROM LESSONS LEARNED

This chapter discussed the building blocks and therefore the need for improved alignment and integration between spatial, transportation and environmental planning and management and the development processes underpinning each discipline, the policy and legal framework guiding it, the division of powers, the functions, duties and obligations, and the opportunities created by the transformation of the existing policy and legal framework provided for with the promulgation of SPLUMA (2013). This perspective and challenge exist in all developed as well as developing countries in the need to be transparent, avoid duplication and accountable in support of strategic planning and management processes.

Perhaps the greatest area of concern is the processes and conditions related to obtaining land-use rights, development planning considerations, project approval in terms of transportation plans inclusive of traffic impact studies and environmental authorizations. All of these processes impact on optimization of spatial planning based on different strategic planning instruments and synchronizing land use activities.

The complexity of current legislative processes and procedures (as applicable to most developing countries), notwithstanding process transformation provided for by SPLUMA (2013) in the case of South Africa, may cause continued uncertainty among spheres of government and professionals in terms of planning, implementation and project monitoring.

Issues such as how to align, engage, integrate, coordinate, support and implement the formulation of spatial development frameworks, land-use management and development planning processes need to be addressed through formal capacity-building and intensive formal training among all stakeholders and all spheres of government. The implementation of transformation instruments such as SPLUMA (2013) will, however, add to the administrative and professional demand within already under capacitated municipalities. The same applies to other spheres of government within all spatial systems where development challenges need to be addressed effectively and efficiently. To this reality political intervention by government in some developing countries may complicate the challenge even further.

The previous spatial and land-use management dispensation in South Africa culminated in the fragmentation of land-use decision-making and requirements, due to the national policy and legislative framework that does not appropriately address the division of functions and powers. Accountability and support in sustainability were thus not reflected and articulated in land-use decisions and environmental authorisations (Kidd, 2008: 85–102). SPLUMA (2013) provides a framework to address the issues of accountability and sustainability. Attention should also be paid to transform all core legislation through amendments in order to promote alignment and integration and not to rely on the provisions guiding co-operative government and intergovernmental relations, as provided for in the Constitution (Act 108 of 1996) only. This challenge also relates to position in other related developing countries and in some cases even to some developed countries.

The promulgation of SPLUMA (2013) and its Regulations (2015) in South Africa directs municipal planning by formalising the role of SDFs within all spheres of government and provides for a new system of land-use management and land development within municipalities. It thus applies to all spheres of government and emphasises the role of municipalities as being the most important sphere of government, as it is closest to the people and communities in terms of delivery. An important focus is the promotion of sustainability in terms of spatial planning, land-use management and land development through alignment and integration.

SPLUMA (2013) thus serves as the interface in promoting alignment and integration within the existing policy and legislative framework guiding planning and development. At this stage, many practitioners view the NDP (2012) as being an *ideology* that needs the support of instruments such as SDFs within all spheres of government in order to determine what is to happen, where and when. SPLUMA (2013) may assist in this, as it serves as important vehicle for alignment, integration and cooperative governance as a prerequisite for a successful democracy.

Contesting spatial planning, land-use and development-planning issues through the Constitutional Court and/or any Court of Law will result in failure and continuation of the past misalignment, lack of integration, and absence of coordination, resulting in continuing spatial planning inefficiency and non-delivery. This specifically applies to the position in most developing countries.

REFERENCES

[1] Republic of South Africa (RSA), *Constitution of the Republic of South Africa, Act 108 of 1996*, Government Printer: Pretoria, 1996.

[2] Geerlings, H. & Stead, D., The integration of land-use planning, transport and environment in European policy and research. *Transport Policy*, **10**(3), pp. 187–196, 2003.

[3] Stead, D., Geerlings, H. & Meijers, E., Integrated land-use planning and environmental policy: An international comparison, Volume 1. OTB Research Institute for the Built Environment, Delft University for Technology: Rotterdam, 2003.

[4] May, A.D., Jopson, A.F. & Mathews, B., Research challenges in urban transport policy. *Transport Policy*, **10**(3), pp. 157–164, 2003.

[5] Republic of South Africa (RSA), *White Paper on Spatial Planning and Land-Use Management*, Government Printer: Pretoria (refer also to the Drafts for 2003, 2006 and 2007), 2001.

[6] Claasens, P.E., The role of environmental management (spatial planning, development planning and conservation of the natural environment) in prompting sustainable development. Report to the National Research Foundation, Pretoria, 2003.

[7] Sowman, M. & Brown, A.L., Mainstreaming environmental sustainability into South Africa's integrated development planning process. *Journal for Environmental Planning and Management*, **49**(5), pp. 695–712, 2007.

[8] Berrisford, S. & Kihato, M., Local government planning legal frameworks and regulatory tools: Vital signs? *Consolidating Developmental Local Government. Lessons from the South African Experience*, eds M. Van Donk, M. Swilling, E. Pieterse & E. Parnell, ISANDLA Institute, UCT Press: Cape Town, pp. 51–75, 2008.

[9] Sowman, M., Integrating environmental sustainability issues into local government planning and decision-making processes. *Democratising Local Government: The South African Experiment*, eds S. Parnell, E. Pieterse, M. Swilling & D. Wooldridge, University of Cape Town Press, Cape Town, pp. 181–203, 2002.

[10] Todes, A., Regional planning and sustainability: Limits and potentials of South Africa's integrated development plans. *Journal of Environmental Planning and Management*, **47**(6), pp. 843–861, 2004.

[11] Todes, A., Sim, V. & Sutherland, C., The relationship between planning and environmental management in South Africa: The case of KwaZulu-Natal. *Planning Practice and Research*, **24**(4), pp. 411–433, 2009.

[12] Schoeman, C.B., Urban and regional planning and the interface with environmental management and transportation planning. Inaugural address. No. 238, Scientific Contributions, North-West University, Potchefstroom Campus, 2010.

[13] Vigar, A., Healey, P., Hull, A. & Davodi, S., *Planning Governance and Spatial Strategy in Britain. An Institutional Analysis*, Macmillan: London, 2000.

[14] Retief, F. & Rossouw, N., Urban environmental policy in Africa. *International Handbook of Urban Policy, Volume 1: Contentious Global Issues*, ed. H.S. Geyer, Edwin Elgar Publishing Limited: Cheltenham, pp. 288–306, 2007.

[15] Du Plessis, D.J. A critical reflection on urban spatial planning practices in post-apartheid South Africa. *Urban Forum*, **25**(1), pp. 68–88, 2014.

[16] Waddell, P., Modelling urban development for land use, transportation, and environmental planning. *Journal of the American Planning Association*, **68**(3), pp. 297–314, 2002.

[17] Department of Local Government and Gtz, *IDP Guide Pack. Guide III-Methodology*, Departmental Publication, Pretoria, 2002.

[18] Republic of South Africa (RSA), Spatial Planning and Land-Use Management (SPLUMA) Act, Act 16 of 2013, Government Printer: Pretoria, 2013.

[19] Republic of South Africa (RSA), Regulations in Terms of the Spatial Planning and Land-Use Management (SPLUMA) Act (16 of 2013)/(R.239), 23 March 2015, Government Printer: Pretoria, 2015.

[20] National Department of Transport. National Transport Master Plan. Departmental publication, Pretoria, 2011.

[21] National Planning Commission (NPC), *National Development Plan*. NPC. Internal Publication, Pretoria. Online. www.gov.za/issues/national-development-plan-2030, 2012.

[22] Department of Cooperative Governance and Traditional Affairs (COGTA), Draft integrated urban development framework, Pretoria, 2014.

[23] National Department of Transport, *Draft National Transport Strategic Framework*, Departmental Publication, Pretoria, 2014.

[24] United Nations Habitat (UN HABITAT), Planning and design for sustainable urban mobility. Global Report on Human Settlements 2013, Nairobi, Kenya, 2013.

[25] United Nations Habitat (UN HABITAT), Towards the development of international guidelines on urban and territorial planning. Nairobi, Kenya, 2014.

[26] United Nations Habitat (UN HABITAT), The state of African cities. Re-imagining sustainable urban transitions. Nairobi, Kenya, 2014.

[27] United Nations Habitat (UN HABITAT), Habitat III: The future we want: The city we need. Nairobi, Kenya, 2014.

[28] Scheepers, T., *A Practical Guide to Law and Development in South Africa. An Introduction to the Law Applicable to the Development Management Process in South Africa*, Juta: Kenwyn, 2000.

[29] Van Wyk, J., *Planning Law*, 2nd ed., Juta: Cape Town, 2012.

[30] Republic of South Africa (RSA), *Development Facilitation Act, Act 67 of 1995*. Government Printer: Pretoria, 1995.

[31] Dewar, D., Todes, V. & Watson, V., Industrial decentralization policy as a mechanism for regional development in South Africa; its premises and record. Urban Problems Research Unit: Project Report. University of Cape Town, Cape Town, 1984.

[32] Tomlinson, R., *Urbanization in Post-Apartheid South Africa*, Unwin Hyman: London, 1990.

[33] Swilling, M., Introduction. *Apartheid City in Transition*, eds M. Swilling, R. Humphries & K. Shubane, Oxford University Press: Cape Town, p. viii, 1991.

[34] Mabin, A. & Smit, D., Reconstructing South African cities? The making of urban planning 1900–2000. *Planning Perspectives*, **12**(2), pp. 193–223, 1997.

[35] PlanAct, *Integrated Development Planning – A Manual for Community Leaders*, Aloe Publications: Braamfontein, Johannesburg, 1997.

[36] Tilman, H. & Wall, R., Urbanization and housing in South Africa. *Housing Generator – Integration or Fragmentation*, ed. H. Tilman, NAI Publishers, e-book: Rotterdam, 1997.

[37] Harrison, P., Huchzermeyer, M. & Mayekiso, M. (eds), *Confronting Fragmentation: Housing and Urban Development in a Democratising Society*, University of Cape Town Press: Cape Town, 2003.

[38] National Planning Commission (NPC), *Diagnostic Overview*. NPC Pretoria, Online. www.gov.za/sites/www.gov.za, 2011.

[39] Oranje, M. & Merrifield, A., National spatial planning in South Africa 1930–2010: An introductory comparative analysis. *Town and Regional Planning*, **56**, pp. 29–45, 2010.

[40] Drewes, J.E. & Van Aswegen, M., National planning in South Africa: A temporal perspective. *Town and Regional Planning*, **62**(2013), pp. 21–28, 2013.

[41] Republic of South Africa (RSA), *Local Government: Municipal Structures Act, Act 117 of 1998*, Government Printer: Pretoria, 1998.

[42] Republic of South Africa (RSA), *Local Government: Municipal Systems Act, Act 32 of 2000*, Government Printer: Pretoria, 2000.

[43] Republic of South Africa (RSA), *Urban Transport Act, Act 78 of 1977*, Government Printer: Pretoria, 1977.

[44] Republic of South Africa (RSA), *National Land Transport Transition Act, Act 22 of 2002*, Government Printer: Pretoria, 2000.

[45] Republic of South Africa (RSA), *National Land Transport Act, Act 5 of 2009*, Government Printer: Pretoria, 2009.

[46] Republic of South Africa (RSA), *R.1208 National Land Transport Act (5/2009): National Land Transport Regulations*, Government Printer: Pretoria, 2009.

[47] Republic of South Africa (RSA), *R.877 National Land Transport Act (5/2009): National Land Transport Regulations on Contracting for Public Transport Services*, Government Printer: Pretoria, 2009.
[48] Republic of South Africa (RSA), *White Paper on National Transport Policy*, Government Printer: Pretoria, 1996.
[49] Republic of South Africa (RSA), *Moving South Africa – The Action Agenda*, Departmental Publication: Pretoria, 1999.
[50] Republic of South Africa (RSA), *Rural Transport Strategy for South Africa*, Departmental Publication: Pretoria, 2003.
[51] National Department of Transport, *Draft NATMAP 2050 Synopsis Report*. Unpublished Departmental Report, Pretoria, 2015.
[52] Republic of South Africa (RSA), *Environment Conservation Act, Act 73 of 1989*, Government Printer: Pretoria, 1989.
[53] Sowman, R., Fuggle, R. & Preston, G., A review of the evolution of environmental evaluation procedures in South Africa. *Environmental Impact Assessment Review*, **15**, pp. 45–67, 1995.
[54] Council for the Environment. *Integrated Environmental Management in South Africa*, Joan Lotter Publications: Pretoria, 1989.
[55] Republic of South Africa (RSA), *National Environmental Management Act, Act 107 of 1998*, Government Printer: Pretoria, 1998.
[56] Republic of South Africa (RSA), *Regulations in terms of Chapter 5 of the National Environmental Management Act, 1998. Regulations R.543 to R.547*, Government Printer: Pretoria, 2010.
[57] Department for Environmental Affairs and Tourism, *A National Framework for Sustainable Development in South Africa*, Departmental Publication: Pretoria, 2008.
[58] Constitutional Court Ruling, City of Johannesburg Metropolitan Municipality (Applicant), Gauteng Development Tribunal (First Respondent), Gauteng Development Appeal Tribunal (Second Respondent) and Ivory-Palms Properties 20 CC (Third Respondent), 2010.
[59] National Treasury, Departmental Publication 2000/01-2006/07, Pretoria.
[60] University of the Western Cape, Community Law Centre, Developmental local government: Determining appropriate functions and powers. Local Government Project, Cape Town, 2007.
[61] Department of Cooperative Governance and Traditional Affairs, *Divisions of Functions*. Departmental Publication: Pretoria, 2002.
[62] Van Wyk, J., Planning in all its (dis)guises: Spheres of government, functional areas and authority. *PER/PLJ*, **15**(5), pp. 288–318, 2012.
[63] Malan, L., Intergovernmental relations and co-operative government in South Africa: The ten-year review. *Politea*, **24**(2), pp. 226–243, 2005.
[64] South African Cities Network, SPLUMA as a tool for spatial transformation, Braamfontein, sacitiesnetwork.co.za, 2015.
[65] Berrisford, S. & De Visser, J., UCT CPD Programme: Lectures on preparing for SPLUMA implementation in co-operation with MCA planners, Cape Town, 2015.
[66] Republic of South Africa (RSA), *Final Draft Land-Use Management Bill*, Government Printer: Pretoria, 2008.
[67] Stead, D. & Meijers, E., Policy integration in practice: Some experiences of integrating transport, land-use planning and environmental policies in local government. *Berlin Conference on the Human Dimensions of Global Environmental Change: Greening of Policies-Linkages and Policy Integration*, Conference Proceedings, 2004.

[68] Department of Provincial Local Government, *Integrated Development Plan (IDP) Format Guide*, Departmental Publication: Pretoria, 2008.
[69] Albrecht, L., Healey, P., & Kunzman, R., Strategic planning and regional governance in Europe. *APA*, **69**(2), 2003.
[70] Albrecht, L., Strategic (spatial) planning re-examined. *Environmental and Planning B; Planning and Design*, 31, pp. 743–758, 2004.
[71] Olesen, K., The neoliberalisation of strategic spatial planning. *Planning Theory*, **13**(3), 2014.
[72] United Nations Environment Programme (UNEP), Integrating the environment in urban planning and management-key principles and approaches for cities in the 21st Century, Nairobi, Kenya, 2013.
[73] Iacono, M., Levinson, D. & El-Geneidy, A., Models of transportation and land use change: A guide to the territory. *Journal of Planning Literature*, SAGE Publications, 2008.
[74] Sivakumar, A., *Modelling Transport: A Synthesis of Transport Modelling Methodologies*, Imperial College London: London, 2007.
[75] Giuliano, G., New directions for understanding transportation and land use. *Environment and Planning A*, **21**, pp. 145–159, 1989.
[76] Eggenberger, M. & Partiario, R., Development of a framework to assist the integration of environmental, social and economic issues in spatial planning. *Impact Assessment and Project Appraisal*, **18**(3), pp. 201–207, 2000.
[77] Goodstadt, V. & Partidaro, M.R., *Spatial Planning and Environmental Assessments. Department of Environmental Affairs. Part III: The Practice.* TEEB for local and regional decision makers. https://www.environment.gov.za/.../economicsof_ecosystems_spatialplanning_enviroassessment.pdf, pp. 105–123, 2010.
[78] Davis, M., *Green Infrastructure In-Depth Case Analysis: Theme 7: Mapping for Planning.* IEEP, Ecologic, GHK, TAU, University of Antwerp, VITO, 2010.
[79] Waddell, P., Integrated land use and transportation planning and modelling: Addressing the challenges in research and practice. *Transport Reviews*, **31**(2), pp. 209–229, 2011.

ENVIRONMENTAL MANAGEMENT AND LAND-USE MANAGEMENT IN SOUTH AFRICA: COMPLEXITIES, CHALLENGES AND OPPORTUNITIES FOR INTEGRATED STRATEGIC PLANNING IN DEVELOPING COUNTRIES

CAREL BENJAMIN SCHOEMAN, DIRK CILLIERS & FRANCOIS RETIEF
Unit for Environmental Sciences and Management, North West University (Potchefstroom Campus), South Africa

ABSTRACT

The interface between environmental management and land-use management forms the focal point to attain sustainable development and resilience within spatial systems. The principle of sustainable development is not a new phenomenon and is being applied by professional practitioners in theory but often not understood in formulation of planning instruments and its application as to promote resilience as a common goal and objective. Environmental practitioners and planning professionals have realized that there is a delicate balance between environments, human, economic, institutional, transportation and spatial planning in articulation of planning, development and management. This balance is however negatively being impacted upon, due to the collapse of this relationship and trans-disciplinary approaches and practices.

The purpose of this chapter is to report on the historic development of the environmental management domain linked to the transformation in spatial planning and land-use management systems from an integration perspective within democratized South Africa and the lessons it holds for developing countries and spatial systems that have emerged from colonialized countries.

The timelines in development and transformation between environmental and spatial planning differ in focus, role and functional impact as far as statutory requirements and processes are concerned. This interface is however most important in all planning cycles applied within spatial systems.

In this chapter an assessment is made of the process in terms of its evolution and the need to align its processes in the enhancement of sustainability and promotion of resilience. The enhancement of this interface through application of strategic planning methodologies, processes, procedures and planning tools are being discussed to deal with environmental management and spatial planning complexities, challenges and opportunities.

The conclusions consist of recommendations to grow and built this interface. It includes improved integration of the policy and legislative frameworks; capacity building amongst the different professions; improved alignment in education, training and professional practice and integrated simulation and modelling techniques, methodologies and the need to formulate trans-disciplinary plans as to promote the integration of the interface.
Keywords: environmental management, spatial planning, land-use systems, trans-disciplinary planning, sustainability and resilience, environmental modelling.

1 INTRODUCTION

This chapter builds on recent work by Retief and Cilliers [1] and Cilliers and Retief [2] on land-use management (LUM) and planning and environmental planning tools (e.g. Environmental Management Framework, EMFs) in South Africa. In this context, LUM is understood as the establishment and implementation of mechanisms to regulate land-use rights through the use of spatial planning instruments, land-use schemes and land development procedures. A distinction is made between 'spatial planning', 'development planning' (dealt with in Section 2) and 'land-use management' (LUM), and the implications it holds for strategic planning and modelling.

During the Apartheid era, spatial planning and development inclusive of land-use management relating to the differentiated and segregated spatial systems and areas for functional planning, statutory control and management [3] relied on a very complex regulatory system. Many of the discriminatory effects of Apartheid in terms of land use and land-use rights were given effect through the LUM system of the time.

Since the new dispensation in 1994 attempts have been made to repeal the plethora of old legislation related to LUM. The aim was to draft a single national act to repeal the previous LUM system. However, progress was slow, with almost a decade passing before the promulgation of the first version of the then 'new' Land Use Bill in 2001 [4], [5], and another decade before SPLUMA was finally enacted in 2013 [6]. The SPLUMA Regulations were published in May, 2015 [7].

A number of 'draft' provincial spatial planning and land-use management acts have also been adopted or are being drafted. Unfortunately, notwithstanding the progress made, the status of land-use rights in some provincial spatial systems are still determined and managed by mechanisms established under legislation from the previous dispensation, some acts dating back to the 1960s. Tables 1 and 2 show the core policies and legislative framework guiding the interface between spatial planning, land-use management and environmental management. From the tables, it is concluded that the LUM system in South Africa remains to be complex and inefficiently governed. However, some standardization and, some would argue, rationality is brought to the equation with the introduction of SPLUMA [6] in 2013.

Parallel to the evolution of the LUM system environmental management has, since 1994, developed a number of innovative environmental planning and assessment tools such as environmental management frameworks (EMFs), strategic environmental assessment (SEA), conservation planning (C-Planning), etc. Furthermore, project level environmental management tools such as environmental impact assessment (EIA), social impact assessment (SIA), heritage impact assessment (HIA), etc. has been developed. In light of these developments it makes sense to critically consider the challenges and opportunities to align and/or integrate the environmental management and LUM systems.

The next section introduces challenges facing of the current LUM system and the implication of the transformation as provided for by SPLUMA [6] after which more specific environmental management tools are discussed in relation to LUM.

2 CHALLENGES FOR LAND-USE MANAGEMENT

The main challenges for LUM in South Africa relate, amongst other to the inherent uncertainties and complexities of the system, which are discussed in this section. Admittedly the new legislative context around the SPLUMA aims to address these challenges, which suggest that we are amidst a transition phase towards more certainty and clarity.

2.1 The status quo and land-use management reality

The land-use management reality in South Africa poses various challenges, specifically related to the uncertainties and complexities inherent to the historic system. For our purpose we identify four key challenges related to the LUM system, building and expanding on the three challenges highlighted by Retief and Cilliers (2015) [1]:

- ***Policy uncertainty (Table 1)**:* This refers to the policies and guidelines that deal with strategic focuses, planning processes and related spatial system realities and focuses.
- ***Legislative uncertainty (Table 2)**:* This refers to the legislative framework which deals with LUM. These acts date from the 1960s to as recent as 2015. The supporting legal

Table 1: Core policies guiding the interface between spatial planning and environmental management. *(Source: author's own compilation, 2015.)*

Land-use management	Environmental management
Reconstruction and Development Plan (1994)	Global Biodiversity Strategy: Guidelines for action to save, study and use earth's biotic wealth sustainably and equitably (published by the WRI, IUCN and UNEP in 1992)
Growth, Economic and Redistribution Strategy (1996)	Balancing the Scales: Guidelines for increasing Biodiversity's Chances through Bioregional Management (published by the World Resources Institute in 1996).
White Paper on South African Land Policy (1997)	Minimum requirements for the Classification, Handling and Disposal of Hazardous Waste (Second Edition) (1998) (DWAF)
Urban Development Framework (1997)	Integrated Environmental Management Guidelines Series (1992)
Rural Development Framework (1997)	DEAT: An Environmental Policy for South Africa (Green Paper) (1996)
White Paper on Water and Sanitation (1997)	White Paper on Integrated Pollution and Waste Management for South Africa (2000)
White Paper on Local Government (1998)	Strategic Environmental Assessment in South Africa (2000)
Green Paper on Development and Planning (1999)	DEAT (2002a) *Screening, Information Series 1,* Department of Environmental Affairs and Tourism (DEAT), Pretoria.
National Integrated Rural Development Strategy (2000)	DEAT (2002b) *Scoping, Integrated Environmental Management,* Information Series 2, Department of Environmental Affairs and Tourism (DEAT), Pretoria.
King Report II on Corporate Governance for South Africa (2002)	DEAT (2002c) *Specialist Studies, Information Series 4,* Department of Environmental Affairs and Tourism (DEAT), Pretoria.
White Paper on Spatial Planning and Land Use Management (2001)	DEAT (2002d) *Impact Significance,* Integrated Environmental Management, Information Series 5, Department of Environmental Affairs and Tourism (DEAT), Pretoria.
National Spatial Development Perspective (2003)	DEAT (2004a) *Overview of Integrated Environmental Management,* Integrated Environmental Management, Information Series 0, Department of Environmental Affairs and Tourism (DEAT), Pretoria.
Mining Charter (2003)	DEAT (2004b) *Criteria for determining Alternatives in EIA*, Integrated Environmental Management, Information Series 11, Department of Environmental Affairs and Tourism (DEAT), Pretoria.
Construction Charter (2005)	DEAT (2004c) *Environmental Impact Reporting,* Integrated Environmental Management, Information Series 15, Department of Environmental Affairs and Tourism (DEAT), Pretoria.
ASGISA (2006)	Strengthening Sustainability in the Integrated Development Planning Process (2001)
National Spatial Development Perspective (2006)	State of the Environment Reporting: Draft Guidelines for Local Municipalities (2005)
Housing Atlas (2006)	DEAT Information Series (2004-2009)
Sustainable Human Settlement Planning: Resource Book (2008) (NDoH)	National Framework for Sustainable Development (2008)
Area-based Planning. Department of Rural Development and Land Reform (2008/2009)	
NPC: A Guide to the National Planning Framework (2009)	
NPC: Green Paper. National Strategic Planning (2009)	
Comprehensive Rural Development Programme. Department of Rural Development and Land Reform (2009)	
National Planning Commission Diagnostic Overview (2011)*	
National Development Plan (2012)*	
Draft Integrated Urban Development Framework (2015)	

*Cross-cutting policies.

Table 2: Core legislative framework guiding interface between spatial planning, land-use and environmental management. *(Source: Author's own compilation, 2015.)*

Spatial planning and land-use management	Environmental management
National Building Regulations and Building Standards Act 103 of 1977	Health Act 63 of 1977
Town Planning and Township Ordinance, Ordinance 15 of 1986	Water Act 54 of 1956
Land Use Ordinance (Cape of Good Hope), Ordinance 15 of 1985	National Water Act 36 of 1991
Removal of Restrictions Act 84 of 1967	Water Services Act 108 of 1997
The Less Formal Township Establishment Act 113 of 1991	National Environmental Management Act 107 of 1998 (NEMA)
The Physical Planning Act 88 of 1967 (Sections 6, 8 and 11)	National Environmental Management: Air Quality Act 39 of 2004
Development Facilitation Act 67 of 1995 (DFA)	National Environmental Management: Waste Act 59 of 2009
Constitution of the Republic of South Africa 108 of 1996	National Environmental Management: Biodiversity Act 10 of 2004 (NEMBA)
Bill of Human Rights 1996	National Environmental Management: Protected Areas Act 57 of 2003 (NEMPAA)
Physical Planning Act 88 of 1967	National Heritage Resources Act 25 of 1999 (NHRA)
Municipal Structures Act 117 of 1998	Mineral and Petroleum Resources Development Act 28 of 2002 (MPRDA)
Restitution of Land Rights Act 22 of 1993	World Heritage Convention Act 49 of 1999
Interim Protection of Informal Rights Act 76 of 1995	Biodiversity Act 10 of 2004
Prevention of Illegal Eviction from Unlawful Occupation of Land Act 19 of 1998	R.543: National Environmental Management Act 107 of 1998): Environmental Impact Assessment Regulations, 2010 (33306)
Reconstruction and Development Programme Act 79 of 1998	R.544: Listing Notice 1: List of activities and competent authorities identified in terms of sections 24(2) and 24D (33306)
Municipal Systems Act 32 of 2000	R.545: Listing Notice 2: List of activities and competent authorities identified in terms of sections 24(2) and 24D (33306)
Development Facilitation Act 67 of 1995 (DFA)	R.546: Listing Notice 3: List of activities and competent authorities identified in terms of sections 24(2) and 24D (33306)
Physical Planning Act 125 of 1991	R.547: Environmental Management Framework Regulations, 2010
Less Formal Township Establishment Act 113 of 1991 (LEFTEA)	
Subdivision of Agricultural Land Act 70 of 1970 (SALA)	
Removal of Restrictions Act 84 of 1967	
Community Land Reform Act 28 of 1996 (CLARA)	
Housing Act 107 of 1997	
National Land Use Management Bill (Draft 2008)	
Local Government: Municipal Integrated Development Planning Regulations, 2001.	
Spatial Planning and Land Use Management Act (SPLUMA) 16 of 2013*	
R. 239 SPLUMA Regulations (23 March 2015)*	

*Cross-cutting legislation.

**Various by-laws exist within municipalities.

principles underpinning the frameworks are included in Scheepers [8] and Van Wyk [9]. From the work by Retief and Cilliers [1] it is also possible to explore the evolution of legislation over time in the following sections.

- ***Institutional and procedural complexity***: The complexity related to institutional arrangements is a result of mandates assigned by legislation, some of which in a highly unconstitutional manner; especially where provincial governments make decisions on municipal planning matters, while the latter is clearly the mandate of a municipal council (as per the Constitutional Court Ruling of 18 June 2010 [10]), where the Court, amongst other, decided that the Constitution envisage a degree of autonomy for the local sphere of government in which municipalities exercise their original constitutional powers free from undue interference from other spheres of government). Retief and Cilliers [1] illustrate this complexity by comparing the different types of land-use management authorisations processes against the relevant legislation and designated authorising authority.
- ***Spatial complexity***: The complicated LUM system is also represented spatially, with different locations administered by different historic legislation, sometimes even more than one act on a given location. This is best illustrated by using a specific local municipality case example from Retief and Cilliers [1] and Cilliers and Retief [2], discussed in the following sections.

2.1.1 Policy uncertainty guiding spatial planning and land-use management

The policy complexities guiding spatial planning and land-use management are shown in Table 1. The United Nations in 2008 [11] states that spatial planning is largely a public-sector function to influence the future spatial distribution of activities (inclusive of land uses). Spatial planning and land-use management thus aims to create a more rational territorial organization of land and the linkages between them, to balance demands for development with the need to protect the environment, and to achieve social and economic objectives.

Spatial planning (inclusive of land-use management) thus consists of measures to coordinate and improve the spatial impacts of other sectoral policies (Table 1). This forms part of the endeavour to achieve a more even distribution of economic development within a given territory than would otherwise be created by market forces. The United Nations report depicts that spatial planning (inclusive of land-use management) forms an important lever for promoting sustainable development and improving the quality of life in all spatial systems.

The report by the United Nations on spatial planning as key instrument for development of effective government includes land-use regulation or development management on municipal level [11]. All policy instruments guiding land-use management should thus be related to spatial planning tools such as spatial development frameworks. At the same time, it should be noted that the formulation of policies may precede, guide or follow the legislative process directing spatial planning and land-use management.

Table 1 demonstrates also the core policy complexities guiding spatial planning and land-use management and its relationship to environmental management. The policies as identified impacts on the formulation of spatial planning tools, land-use management systems in a wider sense regulating land development, potential and control.

2.1.2 Legislative uncertainty guiding spatial planning and land-use management

Table 2 shows the core legislative framework guiding the interface between spatial planning, land-use management and environmental management, while Table 3 provides a legislative analysis over time. Three distinct periods are identified namely, pre-1994, the transition period between 1994 and 2000 and the post-2000 period. The analysis undertaken by Retief and Cilliers [1] illustrates which acts have been repealed and which acts are still in force. The high number of acts reflected in Table 3 supports the notion of legislative uncertainty and complexity as a particular challenge for LUM.

Tracing the origins of LUM (pre-1994 period) in South Africa takes us back to the first Dutch colonization of the Cape where a rudimentary system of land allocation through title deeds were introduced. This was followed by the more stringent LUM system developed in England during the industrial revolution and exported to its colonies in the 1800s (see Van Wyk [12]). Much has been written on the evolution of land management and ownership from the post British colonial era to the formalized Apartheid system in the 1900s. The most prominent legislation during this time was, amongst other, the Native (Black) Urban Areas Act No 21 of 1923 Native (Urban Areas) and Consolidation Act No 25 of 1945 eventually repealed by the Abolition of Influx Control Act No 68 of 1986. These acts had the purpose of regulating the movement and land ownership of individuals according to race. Another set of important legislation was the four provincial town planning ordinances (refer also to Mears [13], Williams [14], Glazewski [15], Wynberg [16]) together with the Group Areas Act, amended 17 times and finally repealed by the Abolition of Racially Based Land Measures Act, No. 108 of 1991. However, during the 1980s the Black Communities Development Act 4 of 1984, and especially Annexure F to this act, served to provide nationally standardized zoning descriptions for so-called '*black areas*'.

Directly following the first democratic election in 1994 there was a concerted effort to do away with the racially based LUM policy and legislation. The two most important new Acts in relation to LUM were Local Government Transition Act (LGTA) 209 of 1993 and Development Facilitation Act (DFA) 67 of 1995 [17]. The first relates to the restructuring of local government and the second to the implementation of the primary action plan for development of the new ANC government, called the reconstruction and development programme (RDP). Both these acts were so-called interim acts to bridge the pre- and post-apartheid eras. Almost immediately after promulgation of the DFA a committee named the Development and Planning Commission was appointed to investigate the complex nature of the LUM system at the time, and to make recommendations for the required legislative reform. Based on the work of the committee a policy Green Paper on Planning and Development [3] and White Paper on Spatial Planning and Land-use Management [18] was published in 1999 and 2001 respectively. During this transition period, various legislation that entrenched 'apartheid' and thus fragmented spatial development were repealed. Transformation of the legislative framework during the post 1994 democratic period still contributed to the legislative complexities guiding LUM.

Ultimately the Green and White papers translated into the first 'new' Land Use Management Bill published in 2001 [18]. Also, based on the work of the Development Planning Commission was the 'new' demarcation of South Africa into 231 Local Municipalities, 47 District Municipalities, 6 Metros and 9 Provinces in an attempt to amongst other literally redraw the LUM map. South Africa now had nine provinces and wall to wall municipalities which meant that the old Ordinances related to the four previous provinces were dysfunctional as well as the implication that local authorities now also decide on land use in rural areas, which previously was the mandate of the National Department of Agriculture [refer also to 19 and 20]. However, it took 12 years and nine revisions of the initial Land Use Management Bill to eventually publish the SPLUMA in 2013 [5]–[7].

Parallel to the national process of drafting SPLUMA, provinces also produced provincial planning legislation, which are now at varying degrees of finality. Indeed, all provinces are at present in process of promulgation of Provincial SPLUMA's based on the principles as included in SPLUMA and its Regulations [6], [7] The fact that the promulgation of SPLUMA is still to be proclaimed, results in the fact that a parallel land-use management system is at present being applied within the national spatial system. Moreover, SPLUMA will on promulgation repeal the following existing legislation (Table 2): Removal of Restrictions Act (84 of 1967); Physical Planning Act (88 of 19670; Less Formal Township Establishment Act (113 of 1991); Physical Planning Act (125 of 1991) and Development Facilitation Act (67 of 1995).

2.2 Institutional and procedural complexity of land-use management

It is not surprising that the inefficient and complex LUM system described in relation to the legislative context in the previous sections produce extreme institutional and procedural complexity. This is best illustrated by the fact that until recently the same application for land-use change could be submitted to different spheres of government and in relation to different Acts. For example, applications for township establishment could be submitted in terms of the DFA or Provincial Ordinances or Less Formal Township Establishment Act. This is, from a decision-making perspective clearly untenable. The 2010 Constitutional Court Ruling [21] which gave clarify on the matter by explicitly confirming that LUM is a local government mandate, and also gave instruction for finalization and promulgation of SPLUMA in 2013 [6] and the repeal of certain sections of the DFA [17].

The complexities and uncertainties around mandate also culminated in complexities in decision making processes, because each act has its own process. Therefore, developers typically cherry pick processes based on their needs, deciding for example if they should apply with local government in terms of Ordinances or the provincial government in terms of the Less Formal Township Establishment Act, 1991 (Table 2). The latter presents a much more streamlined process, but was actually introduced for the purpose of dealing with rapid urbanization and a serious housing need and not for typical township establishment processes.

Table 3 provides an example of the complexities indicative of the institutional and procedural complexities of LUM within a particular province such as the North West Province. However, much of this uncertainty will be addressed by SPLUMA [6], and in future the bypassing of local government in decision making on land-use matters will come to an end.

2.3 Spatial complexity of land-use management

The mandate of different acts is not only related to different spheres of government but also has a specific local spatial dimension. Therefore, certain acts apply to certain geographic areas, as illustrated in Fig. 1 for the Madibeng Local Municipality, which have two Town Planning Schemes (the Brits Town Planning Scheme of 1958 and the Hartebeespoort Town Planning Scheme of 1993) as well as a Peri-Urban Scheme under the Transvaal Town Planning and Townships Ordinance 15 of 1986. Land now covered by a scheme is dealt with by the provincial government in terms of the Physical Planning Act 88 of 1967. Moreover, the municipality also includes tribal land under the control of traditional authorities which falls under the Development Facilitation Act 67 of 1995 [17]. Therefore, the main challenge for municipalities is to find ways of integrating previous land-use management schemes drawn up by different historical acts and ordinances. Various more capacitated local municipalities have already succeeded in integrating their town planning schemes, such as the City of Tshwane and the City of Johannesburg Metropolitan Municipalities.

Table 3: Institutional and procedural complexity of LUM – North West Province example. *(Source: Retief and Cilliers, 2015, pp. 571–573.)*

TYPE OF APPLICATION	DESCRIPTION	RELEVANT LEGISLATION	AUTHORISING AUTHORITY
NATIONAL LEVEL:			
Subdivision of agricultural land	To subdivide agricultural land outside town planning scheme areas.	Act 70 of 1970 – Subdivision of agricultural land	National Department of Agriculture (Approval)
PROVINCIAL LEVEL (North West Province)			
Development area	Applications for township establishment, rezoning, any land-use chances in the former Bophuthatswana areas, situated in North West Province. Any kind of land-use application in any part of North West Province.	Act 67 of 1995 – Development Facilitation Act	Tribunal of the Developmental Facilitation Act. (The Designated Officer and Registrar are officials in Department Developmental Local Government and Housing) (Tribunal Order)
Business rights	Issue permit for any change of land use on agricultural land (former Transvaal)	Section 6 of Act 88 of 1967 – Physical Planning Act	Department Developmental Local Government and Housing (Permit)
Public and/or private resorts	Issue permit for public or private resorts on agricultural land outside town planning scheme areas	Section 6 of Act 88 of 1967 – Physical Planning Act	Department Developmental Local Government and Housing (Permit)
Act 21 of 1940 permission	Issue permit for business rights or resort, if it is stipulated as a restrictive condition in the deed of transfer (Permit) Issue permission for subdivision of agricultural land, if it is stipulated as a restrictive condition in the deed of transfer (Approval/Permission)	Act 21 of 1940 – Advertising on Roads and Ribbon Development Act	Department Developmental Local Government and Housing
Subdivision of agricultural land	Permission to subdivide agricultural land, when the farm is excluded from Act 70 of 1970	Ordinance 20 of 1986 – Division of Land Ordinance	Department Developmental Local Government and Housing (Approval)
Cancellation of certificate	When a holding is needed for township establishment, the holding must be excluded from Act 19 of 1919	Act 22 of 1919 – Agricultural Holdings Act	Department Developmental Local Government and Housing (Certificate of cancellation)
Removal of restrictive conditions	Removal of restrictive conditions from a Deed of Transfer	Act 84 of 1967 – Removal of Restrictions Act	Department Developmental Local Government and Housing (Approval)

Table 3: Continued.

TYPE OF APPLICATION	DESCRIPTION	RELEVANT LEGISLATION	AUTHORISING AUTHORITY
PROVINCIAL LEVEL (North West Province)			
Township establishment	Township establishment where a serious housing need can be motivated	Act 113 of 1991 – Less Formal Township Establishment Act	Department Developmental Local Government and Housing (Approval)
Township establishment	Certain permissions must be giving by the administrator/premier, in terms of the regulations in the township establishment process. (This is only where the original township application was submitted in terms of this regulations – no new township applications was submitted in terms of this regulations)	Regulation Relating to Township Establishment and Land Use, 1986 (Government Notice No. R 1897 of 12 September 1986)	Department Developmental Local Government and Housing (Approval)
Extension of a town planning scheme area or including of a farm portion in a town planning scheme area	Extension of town planning schemes or inclusion of a farm portion in the town planning scheme area	Section 88 of Ordinance 15 of 1986	Department Developmental Local Government and Housing (Approval)
Declaration of municipality as an authorized authority	Declaration of a municipality as an authorized authority in terms of the Town Planning and Township Ordinance; and the Division of Land Ordinance.	Section 2 of the Town Planning and Township Ordinance 1986, Ordinance 15 of 1986. Section 3 of the Division of Land Ordinance 20 of 1986	Department Developmental Local Government and Housing (Approval)
Townships Board	When abjections were received for applications submitted in terms of Physical Planning Act 88 of 1967, Ordinance 20 of 1986, Act 22 of 1919	Ordinance 15 of 1986	Department Developmental Local Government and Housing (The recommendation of the Township Board is submitted to the MEC, the Dept then issue an approval or a permit)

Table 3: Continued.

TYPE OF APPLICATION	DESCRIPTION	RELEVANT LEGISLATION	AUTHORISING AUTHORITY
PROVINCIAL LEVEL (North West Province)			
Public resorts	Establishment of Provincial public resorts	Ordinance 18 of 1969	Provincial Government
Township establishment	Township Establishment in terms of Annexure F	Annexure F of the Black Communities Development Act 4 of 1984	Provincial Government
LOCAL MUNISIPALITY LEVEL			
Township establishment	Township establishment applications in the former Transvaal areas in North West Province	Town Planning and Township Ordinance 1986, Ordinance 15 of 1986 and regulations	Relevant municipality (Approval)
Township establishment	Township establishment applications in the former Northern Cape areas in North West Province	Ordinance 15 of 1985 (Cape of Good Hope) and regulations	Relevant municipality (Approval)
Rezoning	Change of land use stipulated by a town scheme	Ordinance 1986, Ordinance 15 of 1986 and regulations and applicable local town schemes, regulations and by-laws.	Relevant municipality (Approval)
Subdivision of land	Subdivision of erven within the town planning scheme area of local authorities	Ordinance 1986, Ordinance 15 of 1986 and regulations and applicable local town schemes, regulations and by-laws.	Relevant municipality (Approval)
Road closures	Approval of municipal road closures	Local government ordinance 17 of 1939	Local Government
Park closures	Approval of municipal parks closures	Local government ordinance 17 of 1939	Local Government

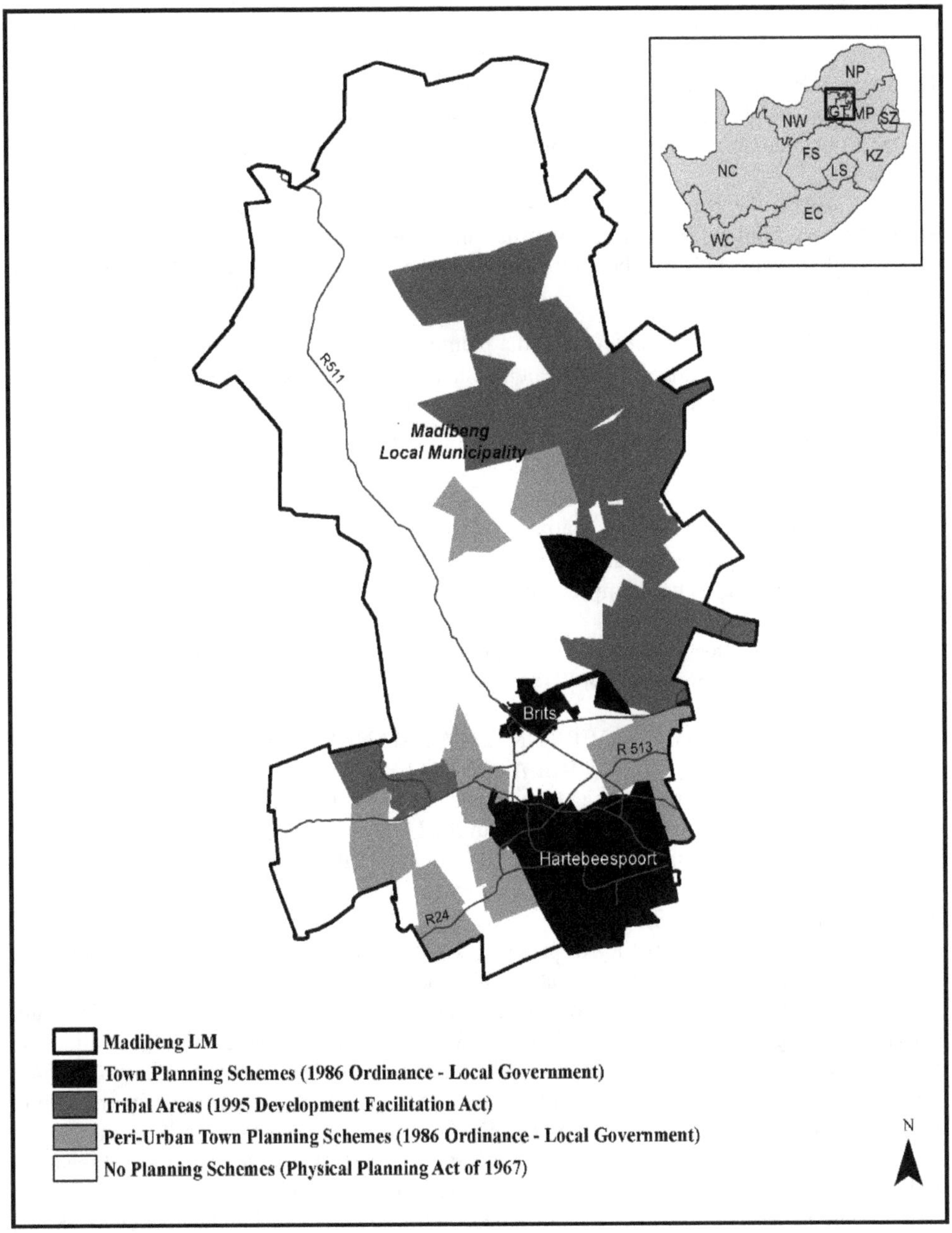

Figure 1: Spatial complexities of the LUM system – Madibeng Local Municipality example. *(Source: Retief and Cilliers, 2015, p. 575 [1].)*

The recent approval of SPLUMA [6] and its Regulations [7] will address most of these problems. As mentioned above SPLUMA and its regulations still need to be promulgated. Each province will also have its own SPLUMA and Regulations. Each municipality will either have a Municipal Planning Tribunal (MPT) or be part of a Joint Municipal Planning Tribunal (JMPT) [6]. Local Municipalities (LM's) will formulate and publish its own policies for a MPT or JMPT. Of importance in SPLUMA [6], [7] is the composition of the tribunals that will not include any political representatives that will enhance the autonomy of such structures.

Fig. 2 shows the benefits of a well-functioning Land Use Management System [22]. SPLUMA as a Tool for Spatial Transformation is well covered in the South African Cities Network [23]. SPLUMA will be addressing the transformation of the LUM domain in the following context:

- SPLUMA defines each element of the planning system from strategic spatial planning to management of land development and its interrelated links between spatial planning, development planning and land-use management.
- It consists of legislative principles planning guiding normative direction to spatial and LUM.
- It addresses based on planning principles fragmented and unsustainable spatial development patterns within spatial systems.
- It creates a single, integrated legal system guiding spatial planning and LUM in a uniform fashion. The complexities as included above are thus addressed in an integrative fashion.
- SPLUMA specifies the role of each sphere of government involved in spatial planning, spatial development and LUM [23].

3 STRATEGIC ENVIRONMENTAL MANAGEMENT TOOLS RELATED TO LUM

Because many environmental management tools engage with strategic planning and have a strong spatial dimension they relate strongly to the LUM system. South Africa has taken the lead in the development and implementation of many of these tools, such as conservation planning (C-Planning), environmental management frameworks (EMFs) [2] and strategic environmental assessment (SEA).

South Africa has a long history of project level environmental impact assessment (EIA), which has a direct relation to applications for land-use change and/or re-zoning. Fig. 3 provides a summary of the extent of strategic environmental management tools in South Africa, while Fig. 4 shows that around 76% of South Africa has been covered spatially. The main message is that local municipalities have taken a strong lead in terms of environmental management tool development. Since LUM is a local mandate this is particularly promising, because it suggest that significant potential exist to integrate LUM and environmental management at local level. In terms of strategic environmental management tools three tools have particular relevance namely, SEA, EMF [2] and C-Planning.

SEA was the earliest strategic tool introduced in South Africa during the mid-1990s. Since then various SEA guidelines have been developed [25]–[27] reflective of the extensive practice that has emerged.

Internationally SEA is basically understood as the assessment of strategic level actions such as policies, plans or programmes [27]. However, in South Africa SEA has been understood as much more of a planning tool as well, not only an assessment tool [28]–[30]. This broader understanding has led to it being aligned well with the planning system.

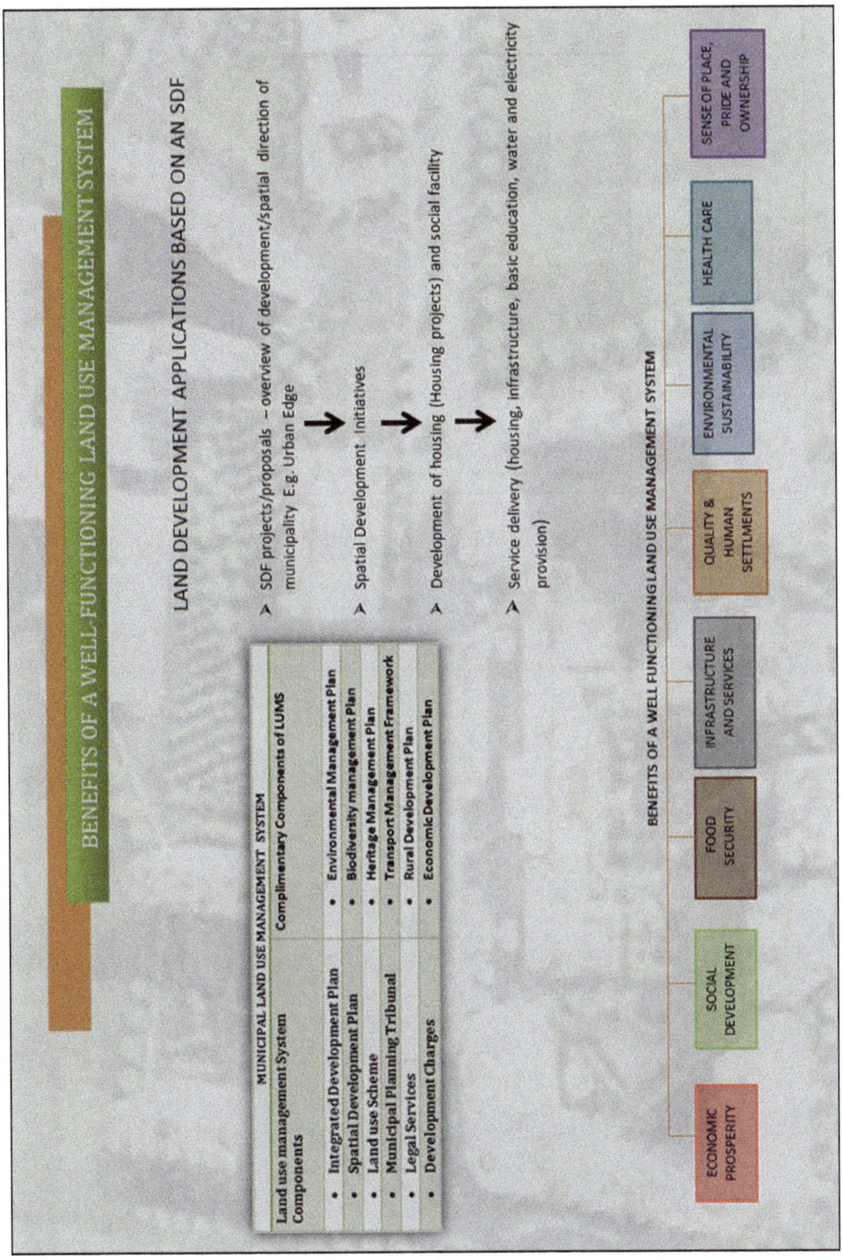

Figure 2: Benefits of a well-functioning LUMS [22].

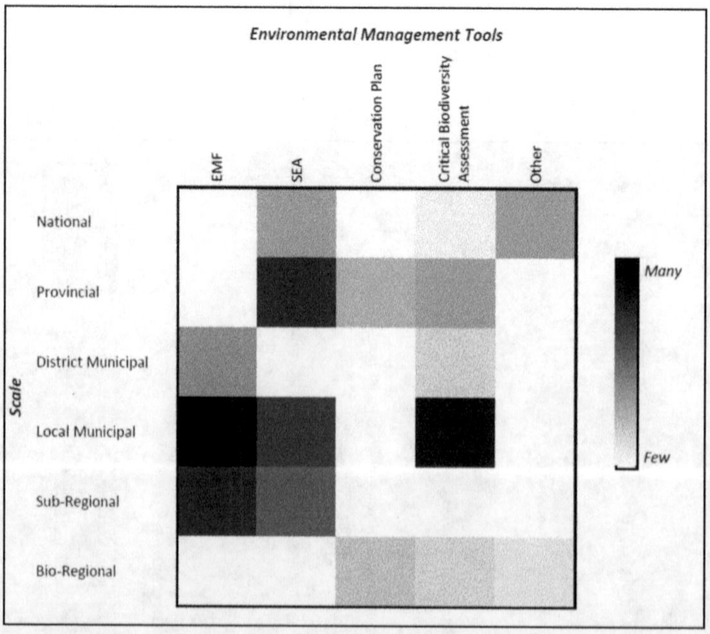

Figure 3: Extent of the application of strategic environmental management tools. *(Source: Retief and Cilliers, 2015, p. 577 [1].)*

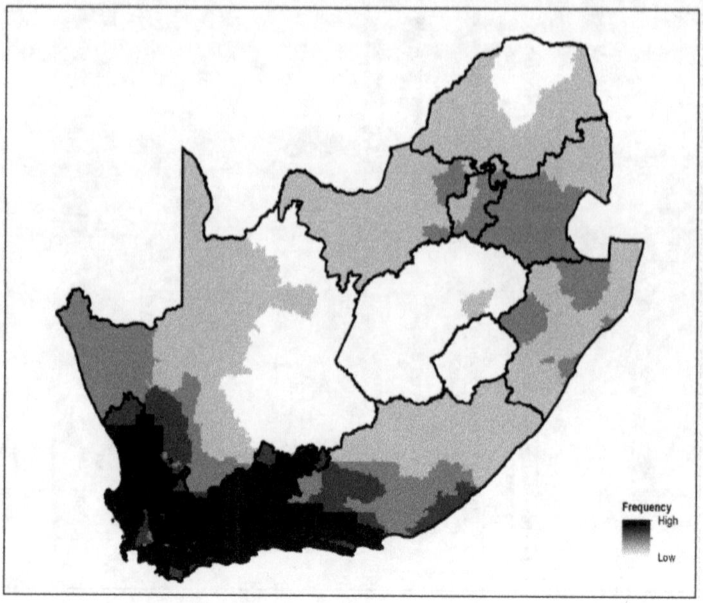

Figure 4: Extent of EMF and C-Planning in South Africa. *(Source: Retief and Cilliers, 2015, p. 576 [1] and Cilliers and Retief [2]. Refer also to Cilliers and Retief [24].)*

In this regard the planning legislation even makes explicit reference to SEA, for example White Paper on Spatial Planning and Land Use Management of 2001 [18] as well as the Municipal Planning and Performance Management Regulations of 2001 [31]. The Spatial Planning and Land Use Management Act (2013) [6] and [7] provides for the inclusion of a strategic assessment of environmental pressures and opportunities within the municipal area including the spatial location of environmental sensitivities. In terms of SEA practice between 1996 and 2003 more than a third of SEAs were planning related. The link between planning and SEA has also been widely discussed and debated in various literature [33]–[36]. The challenges for the SEA in relation to LUM system are defined by Rossouw and Retief [33] and Retief and Cilliers [1] as follows:

- expanding and refining the legislative basis for SEA;
- preparing clear guidance on the integration of SEA with LUM;
- reviewing the SEA system to identify barriers to SEA performance; and
- improving professional capacity to apply and implement SEA.
- SEA should be aligned with nationally and locally defined sustainable development objectives;
- SEA should form an integral part of the planning process to avoid duplication and enhance effectiveness;
- SEA should address the entire planning process by including a management, implementation and monitoring plan;
- SEA can contribute to a decision-making framework by establishing limits of acceptable change, thresholds, sustainability criteria and indicators; and
- SEA should fit the purpose for which it is applied to avoid it becoming an additional burden in an already complex planning process (refer also to Retief [37], [38], DEAT [39], DEA [40]).

EMF is different to SEA in that it enjoys an explicit legal mandate in terms of NEMA and relevant EMF Regulations (GNR 547) with accompanied guidelines [40]. There has been a difference in the interpretation of the exact purpose of EMF but there is general agreement that it should provide a spatial understanding of environmental sensitivity and that this information should inform both spatial planning in the form of SDFs and project level EIA applications. Recent research by Cilliers and Retief [2] shows that municipalities have been the main drivers for EMF development which is surprising since EMF is a provincial and national mandate. One particular area of confusion in terms of the relationship between EMF and LUM mandates is the use of the term 'land use' in the aim of EMF described in the regulations, namely:

"Environmental management framework" means a study of the biophysical and socio-cultural systems of a geographically defined area to reveal where specific land uses may best be practiced and to offer performance standards for maintaining appropriate use of such land.

Since NEMA regulates 'activities' and not 'land use' this is clearly problematic. The national Department of Environmental Affairs has however explained that they used the term land use in referring to the use of land and not land use as in the rights on land. Apart from the seemingly confusing mandate, research on the quality of EMF practice suggests following challenges for EMF in relation to LUM (refer also to Retief [35], Retief et al. [36]).

- Conflicting mandates: The mandate for EMF is vested at provincial level while LUM is essentially a local government function (as confirmed in the Constitutional Court Ruling of 18 June 2010) and SPLUMA [6], [7], [23]. The challenge therefore is to integrate EMF outcomes with SDFs and other LUM policies at the local level as provided for in SPLUMA (2013) [6], [7].

- Defining sensitivity: A common definition and methodology is required to deal with sensitivity in EMFs. This will enable comparison as well as the integration of different EMFs.
- Dealing with public participation: EMFs could potentially benefit greatly by integrating the public participation components with IDP, SDF and LUM processes.

The third strategic environmental planning tool to consider is so-called C Planning or Biodiversity Planning, which is more of a methodological and less policy driven tool. This is because C-Planning is mainly understood as a particular method in conservation biology and not a legal tool per se. The aim of C-Planning is to identifying areas of critical biodiversity, developing a strategy/plan to protect these areas, aligning the strategy with existing planning tools and frameworks and implementing the strategy.

C-Planning is applied though the application of the systematic conservation planning method, which is strongly GIS based [40]–[44]. The method relies strongly on the identification of conservation targets and strategies to achieve these targets. However, due to the lack of an explicit legal mandate questions have been asked in terms of the contribution of these plans to decision making and their actual effectiveness [45]–[50]. Therefore, C-Planning is viewed as mainly an information tool. It is a concern that according to Knight et al. [49] and Knight and Cowling [50]) only one third of the conservation plans (globally) published between 1998 and 2000 were implemented, referred to as the so-called 'research-implementation-gap' (Knight et al. [49], Knight and Cowling [50]). It is understood that the way to improve the effectiveness of C-Planning in terms of influencing decision making is via its integration with the LUM system and specifically SDFs. This has been a particular point made by the South African National Biodiversity Institute (SANBI) in 2004, who highlighted several challenges with regard to its integration with LUM. Basically, conservation targets identified though C-Planning need to be incorporated with the strategies land development objectives/targets contained in local SDFs. However, it remains hard to see how this will be successful without an explicit legal mandate to do so (refer also to Pierce et al. [51], Knight and Cowling [52], Knight et al. [53], Gibson et al. [54]).

Finally, the integration of project level EIA with land-use change decisions has been an ongoing debate in South Africa for almost three decades. Currently it is well understood by the environmental assessment as well as planning professions that EIA and land-use change applications happens in terms of different mandates with decision making vested at different sphere of government. Moreover, it is interpreted that EIA regulates 'activities' described in the EIA regulations and that land-use change relates to 'land use' rights described in title deeds and/or town planning schemes (Land Use Schemes). However, the evolution of EIA from a more conservation focussed tool dealing with the biophysical environment towards a tool geared more towards dealing with sustainability and in the case of South Africa, specifically dealing with section principles described in NEMA, has made its mandate very broad and potentially overlapping with LUM (refer to Wood [55], Retief et al. [56], Kidd and Retief [57], Retief and Kotze [58], Kidd [59]).

Dealing with the integration of EIA and LUM processes is also an area of concern especially in relation to the general aim of government to improve efficiency in decision making. The publication of the Spatial Planning and Land Use Management Act (2013) [6], [7] explicitly stresses the need for process efficiency both in its objectives and normative principles [23]. It states that the objective of the Act is *inter alia* to *"provide for a framework for spatial planning and land-use management........to provide for a framework for policies, norms and standards for spatial planning and land use management...."*

Although EIA has in the past been blamed for delaying development, recent research [60], [61] suggest that LUM authorisations takes significantly longer than environmental authorisation processes (a matter which is to be addressed in the SPLUMA (2013) [6], [7]). This might be because the EIA systems at this stage seem to be less complex, governed by a single Act (excluding mining) while the LUM system is in process of transformation through SPLUMA [6], [7] while the different provincial SPLUMA's is being formalized. At this stage and during the migration to the new spatial planning, development planning and land-use management system, LUM will still remain disjointed (Tables 1, 2 and 3). It is difficult to see how process alignment and integration could be achieved successfully until SPLUMA with its new system has been promulgated.

4 CONCLUSION AND WAY FORWARD

This chapter explains and sets out various challenges to municipalities in migration to the new LUM system as provided for in SPLUMA. It should be noted that there are several challenges that need to be addressed in the alignment and integration of spatial (LUM), transport and environmental management. The integration possibilities that are being provided for relate to strategic and planning alignment and integration and not so much related to process integration especially on project level.

This opportunity is being provided for in SPLUMA [3], [5] with the emphasis on provisions related to:

- The new framework for spatial planning and land-use management applicable to all spatial systems.
- The defined relationship between the spatial planning and land-use management systems and related kind of planning.
- The inclusive, developmental, equitable and effective planning at different spheres of government.
- Provision of a framework for monitoring, coordination and review of the spatial planning and land-use management system.
- Provision of a framework for policies, principles, norms and standards for spatial development planning and land-use management.
- Addressing of past spatial regulatory processes and imbalances.
- Promotion of greater consistency and uniformity in application procedures and decision making by authorities responsible for land-use decisions and development applications.
- Provision for the establishment, functions and operations of Municipal Planning Tribunals.
- Provision for the facilitation and enforcement of land use and development measures.

The abovementioned new focus of LUM, provide for a new and improved interface with environmental management. The development of various related environmental management tools described in the previous section provides excellent opportunity to strengthen the interface between the environmental and LUM, as also promoted by SPLUMA [6], [7], [15], [62], [63]. However, in order to take full advantage of the potential contribution of environmental management to LUM, Retief and Cilliers [1] suggest that policy, procedural and institutional integration will be required, as also argued throughout this chapter. This seems particularly achievable since the principles of NEMA and SPLUMA seems to be well aligned. Moreover, the process and institutional cooperation seems to be improving, especially with the gradual introduction of SPLUMA. However, it will be difficult to fully achieve and optimise the interface between environmental management and LUM until such time as the planning profession has fully migrated to the new LUM system as provided for in SPLUMA.

One of the challenges after the promulgation of SPLUMA (2013) will be the opportunity to align all related policy, legislative and procedural practices between LUM an environmental management [63]. The planning profession is at present through its Professional Body (SACPLAN) in process of finalization of the professionalization of the profession in terms of competencies, standards, RPL, CPD, job profiling and work reservation. Perhaps this interface provides the opportunity to professionalise environmental management further than what is being provided for in terms of the South African Council for Natural Scientific Professions (SACNASP) (2003). However, in terms of the alignment of LUM and environmental management the foundation now exists to formalise this relationship and interface as dealt with above.

REFERENCES

[1] Retief, F. & Cilliers, D., Land use management and planning. *South African Environmental Law: The Governing Function of Local Government*, ed. A. Du Plessis, Juta: Cape Town, pp. 563–586, 2015.

[2] Cilliers, D. & Retief, F., The status and extent of environmental management frameworks (EMFs) in South Africa, 2000–2015. *South African Geographical Journal*, submitted 2016.

[3] National Department of Land Affairs, Green Paper on Development Planning. Government Printers: Pretoria, 1999.

[4] Christopher, A.J., *The Atlas of Apartheid*, Witwatersrand University Press: Johannesburg.

[5] Republic of South Africa (RSA), Final Draft Land-Use Management Bill. Government Printers: Pretoria, 2008.

[6] Republic of South Africa (RSA), Spatial Planning and Land-Use Management (SPLUMA) Act, Act 16 of 2013. Government Printers: Pretoria, 2013.

[7] Republic of South Africa (RSA), Regulations in Terms of the Spatial Planning and Land-Use Management (SPLUMA) Act (16 of 2013)/(R.239), 23 March 2015. Government Printers: Pretoria, 2015.

[8] Scheepers, T., A practical guide to law and development in South Africa. An introduction to the law applicable to the development management process in South Africa. Juta: Kenwyn, 2000.

[9] Van Wyk, J., *Planning Law*, 2nd ed., Juta: Cape Town, 2012.

[10] Constitutional Court Ruling. City of Johannesburg Metropolitan Municipality (Applicant), Gauteng Development Tribunal (First Respondent), Gauteng Development Appeal Tribunal (Second Respondent) and Ivory-Palms Properties 20 CC (Third Respondent), 2010.

[11] United Nations, *Spatial Planning Key Instrument for Development and Effective Governance with Special Reference to Countries in Transition. Economic Commission for Europe*, New York and Geneva, 2008.

[12] Van Wyk, J., *Planning Law: Principles and Procedures of Land Use Management*, Juta: Cape Town, 1999.

[13] Mears, R., Economic factors affecting migration in South Africa. Unpublished Research Paper, University of Johannesburg, 2004.
[14] Williams, J.J., South Africa, urban transformation. *Cities*, **17**, p. 183, 2000.
[15] Glazewski, J., *Environmental Law in South Africa*, Butterworths, 2000.
[16] Wynberg, R., A decade of biodiversity conservation and use in South Africa: Tracking progress from the Rio Earth Summit to the Johannesburg World Summit on sustainable development. *South African Journal of Science*, pp. 233–234, 2002.
[17] Republic of South Africa (RSA), Development Facilitation Act, Act 67 of 1995. Government Printer: Pretoria, 1995.
[18] Republic of South Africa (RSA), White Paper on Spatial Planning and Land Use Management. Government Printer: Pretoria, 2001.
[19] Republic of South Africa (RSA), Local Government: Municipal Structures Act, Act 117 of 1998. Government Printer: Pretoria, 1998.
[20] Republic of South Africa (RSA), Constitution of the Republic of South Africa, Act 108 of 1996. Government Printer: Pretoria, 1996.
[21] Constitutional Court Ruling. City of Johannesburg Metropolitan Municipality (Applicant), Gauteng Development Tribunal (First Respondent), Gauteng Development Appeal Tribunal (Second Respondent) and Ivory-Palms Properties 20 CC (Third Respondent), 2010.
[22] Department of Rural Development and Land Reform. Departmental information on SPLUMA. Pretoria, 2014.
[23] South African Cities Network, *SPLUMA as a Tool for Spatial Transformation*, Braamfontein: Johannesburg, 2015.
[24] Cilliers, D. & Retief, F., The art and science of conservation planning at municipal level. *Biodiversity Planning Forum Conference* (Conference Proceedings), Wilderness, South Africa, 2011.
[25] Retief, F., The evolution of environmental assessment debates – critical perspectives from South Africa. *Journal of Environmental Assessment Policy and Management*, **12**(4), pp. 1–23, 2010.
[26] Retief, F., Jones, C. & Jay, S., The emperor's new clothes – reflections on SEA practice in South Africa. *Environmental Impact Assessment Review*, **28**, pp. 504–514, 2008.
[27] Dalal-Clayton, B. & Sadler, B., *Strategic Environmental Assessment: A Sourcebook and Reference Guide to International Experience*, Earthscan, 2005.
[28] Department of Environmental Affairs and Tourism (DEAT), Strategic Environmental Assessment (SEA) in South Africa: Guideline Document. Government Printer, 2000.
[29] Department of Environmental Affairs and Tourism (DEAT), Strategic Environmental Assessment, Integrated Environmental Management, Information Series 10. Government Printer: Pretoria.
[30] Department of Environmental Affairs and Tourism (DEAT), 2007 Strategic Environmental Assessment, Integrated Environmental Assessment Guideline Series 4. Government Printer: Pretoria.
[31] GN R796 GG 22605 of 24 August 2001, Local Government: Municipal Planning and Performance Management Regulations. Department of Rural Development and Land Reform (DRDLR). Government Printer: Pretoria.
[32] Wiseman, K., Environmental assessment and planning in South Africa: The SEA connection. *Perspectives on SEA*, eds M. Partidario & R. Clark, *CRC Press*, 2008.
[33] Rossouw, N. & Retief, F., Strategic Environmental Assessment and planning in South Africa. *Strategic Environmental Assessment and Land Use Planning: An International Evaluation*, eds C. Jones, M. Baker, S. Jay, M. Short & C. Wood, Earthscan, 2005.

[34] Rossouw, N. & Govender, K., Strategic Environmental Assessment (SEA) and Development Planning in South Africa. *Environmental Assessment Outlook*, **3**, pp. 70–74, 2003.

[35] Retief, F., The evolution of the environmental debates-critical perspectives from South Africa. *Journal for Environmental Assessment Policy and Management*, **12**(4), pp. 1–23, 2010.

[36] Retief, F., Jones, C. & Jay, S., The status and extent of Strategic Environmental Assessment (SEA) practice in South Africa – 1996–2003. *South African Geographic Journal*, **89**(1), pp. 44–54, 2007.

[37] Retief, F., The quality and effectiveness of spatial planning related Strategic Environmental Assessment (SEA) within the South African context: a case study. *Town and Regional Planning*, **52**, pp. 6–19, 2007.

[38] Retief, F., A performance evaluation of strategic environmental assessment (SEA) processes within the South African context. *Environmental Impact Assessment Review*, **27**, pp. 84–100, 2007.

[39] Department of Environmental Affairs and Tourism (DEAT), A national strategy for Integrated Environmental Management in South Africa – Discussion Document. Government Printer, 1998.

[40] Department of Environmental Affairs (DEA), Guideline 6, Environmental Management Frameworks. Government Printer: Pretoria, 2011.

[41] Marais, M., *The quality of Environmental Management Frameworks in South Africa*, unpublished dissertation for Masters in Environmental Management, North West University, 2011.

[42] De Villiers, C. & Hill, R., Environmental Management Frameworks as an alternative to farm-level EIA in a biodiversity hotspot: A proposal from the Cape Floristic Region, South Africa. *Journal of Environmental Assessment, Policy and Management*, **10**(4), pp. 333–360, 2008.

[43] Reyers, B., Ginsburg, A.E., Nel, J.L., O'Farrell, P.J., Roux, D.J. & Cowling, R.M., From conservation biology to conservation science, moving beyond the disciplinary divides. *Exploring Sustainable Science*, eds M. Burns & A. Weaver, African Sun Media, 2008.

[44] Sarkar, S., et al., Biodiversity conservation planning tools: Present status and challenges for the future. *Annual Review of Environmental Resources*, **31**, pp. 123–159, 2006.

[45] Escott, B., Using GIS and Systematic Conservation Planning Software, Jan/Feb Position It, 2011.

[46] Margules, C.R. & Pressey, R.L., Systematic conservation planning. *Nature*, **405**, pp. 243–254, 2000.

[47] Prendergast, J.R., Quinn, R.M. & Lawton, J.H., The gaps between theory and practise in selecting nature reserves. *Conservation Biology*, **13**(3), pp. 484–492, 1999.

[48] Balmford, A. & Cowling, R.M., Fusion or failure? The future of conservation biology. *Conservation Biology*, **20**, pp. 692–695, 2006.

[49] Knight, A., Cowling, R. & Campbell, B., An operational model for implementing conservation action. *Conservation Biology*, **20**(2), pp. 408–419, 2006.

[50] Knight, A. & Cowling, R., Embracing opportunism in the selection priority conservation areas. *Conservation Biology*, **21**, pp. 1124–1126, 2007.

[51] Pierce, S., Cowling, R., Knight, A.T., Lombard, A.T., Rouget, M. & Wolf, T., Systematic conservation planning products for land-use planning: Interpretation for implementation. *Biological Conservation*, **125**(4), pp. 441–458, 2005.

[52] Knight, A. & Cowling, R., Conserving South Africa's 'lost' biome: A framework for securing effective regional conservation planning in the subtropical thicket biome. Terrestrial Ecology Research Unit Report No. 44, Nelson Mandela Metropolitan University: Port Elizabeth, 2003.

[53] Knight, A., Cowling, R., Rouget, M., Balmford, A., Lombard, A. & Campbell, B., Knowing but not doing: selecting priority conservation areas and the research implementation gap. *Conservation Biology*, **22**(3), pp. 610–617, 2008.

[54] Gibson, R., Hassan, S., Holtz, S., Tansey, J. & Whitelaw, G., *Sustainability Assessment Criteria, Processes and Applications*, Earthscan Publications, 2005.

[55] Wood, C., *Environmental Impact Assessment: A Comparative Review*, Prentice Hall, 2003.

[56] Retief, F., Welman, C. & Sandham, L., Performance of Environmental Impact Assessment (EIA) screening in South Africa: A comparative analysis between the 1997 and 2006 EIA regimes. *South African Geographical Journal*, **93**(2), pp. 1–18, 2011.

[57] Kidd, M. & Retief, F., Environmental assessment. *Fuggle and Rabie's Environmental Management in South Africa*, eds H. Strydom & N. King, Juta Publishing, 2009.

[58] Retief, F. & Kotze, L., The lion, ape and the donkey: Cursory observations on the misinterpretation and misrepresentations of environmental impact assessment (EIA) in the chronicles of the Fuel Retailers. *South African Journal of Environmental Law and Policy*, **15**, pp. 138–155, 2008.

[59] Kidd, M., Removing the green-tinted spectacles: The three pillars of sustainable development in South African environmental law. *South African Journal of Environmental Law and Policy*, **15**(1), pp. 85–102, 2008.

[60] Fuel Retailers Association of Southern Africa v Director-General Environmental Management, Department of Agriculture, Conservation and Environment, Mpumalanga Province and Others. SCA, 109, 2006.

[61] Van Schalkwyk, M., Environmental protection: Quicker, simpler, better, http://www.saroof.org.za. 2006. Accessed on: 20 Jul. 2009.

[62] Steenkamp, C., *The efficiency and alignment of planning and environmental impact assessment (EIA) authorisation processes in the Mpumalanga Province*, Unpublished dissertation for Masters in Environmental Management, North West University, 2010.

[63] Schoeman, C.B., The alignment between spatial planning, transportation planning and environmental management within the new spatial systems in South Africa. *Town and Regional Planning*, **67**, 2015.

WIT*PRESS* ...for scientists by scientists

Land Use Management and Transportation Planning

*Edited by: **C.B. SCHOEMAN**, North West University, South Africa*

The interface between land use management and transportation planning represents probably the most important spatial impact in sustainable land use, mobility and transportation development. Prior to this book, only limited attempts have been made to integrate these topics so as to enhance smart growth and sustainable development principles within spatial systems.

The approach followed differs internationally and specifically between different planning and transportation authorities. The spatial impacts of land use and transportation serve as the main catalyst in urban form, development and its associated problems. These impacts represent severe consequences from a built and environmental development perspective. All of these are covered in the book and its supporting chapters.

The focus of the book is the application of best practice principles in managing the interface between land use management and transportation planning. Internationally the practice is the promotion of more sustainable urban and rural forms supported by improved levels of accessibility through the application of smart growth and sustainability principles. The focus however remains to successfully optimise land use and transportation integration.

The structuring used within each of the chapters provides the reader with the basic and applicable theory and practical knowledge to attain system-wide integration and sustainability within the dynamics of spatial and transportation systems. The inclusion of specific theme-related case studies endorses the relevancy of this book's topic.

WIT Transactions on State-of-the-art in Science and Engineering, Vol 86
ISBN: 978-1-78466-077-2 e-ISBN: 978-1-78466-078-9
Published 2015 / 240pp

First International Symposium on Urban Development

Koya as a Case Study

*Edited by: **F.M. KHOSHNAW**, Koya University, Iraq*

This book contains papers presented at the International Symposium on Urban Development held in the Kurdistan region of Iraq, organised by the Faculty of Engineering at Koya University.

The Kurdistan region is rich in oil, gas, mineral resources and underground water. However, until recently the political and security issues were such that the region was unable to take advantage of those resources. Nowadays, Kurdistan is emerging as one of the fastest developing areas in the Middle East, with its universities playing a major role in this process.

The International Symposium papers included in this volume cover a wide range of topics and are written by people with different specialisations and perspectives. They relate the research carried out at academic and government institutions with the needs of Society.

WIT Transactions on State-of-the-art in Science and Engineering, Vol 77
ISBN: 978-1-84564-888-6 e-ISBN: 978-1-84564-889-3
Published 2014 / 376pp

www.ingramcontent.com/pod-product-compliance
Lightning Source LLC
LaVergne TN
LVHW062239200926
848484LV00006B/1575

* 9 7 8 1 7 8 4 6 6 2 3 3 2 *